T0252190

Overhead Distribution Lines

IEEE Press
445 Hoes Lane
Piscataway, NJ 08854

IEEE Press Editorial Board
Ekram Hossain, *Editor in Chief*

Jón Atli Benediktsson	David Alan Grier	Elya B. Joffe
Xiaoou Li	Peter Lian	Andreas Molisch
Saeid Nahavandi	Jeffrey Reed	Diomidis Spinellis
Sarah Spurgeon	Ahmet Murat Tekalp	

Overhead Distribution Lines

Design and Application

Lawrence M. Slavin, PhD

IEEE Standards designations are trademarks of The Institute of Electrical and Electronics Engineers, Incorporated (www.ieee.org/).

Non-IEEE trademarks are the respective property of their owners.

Copyright © 2021 by The Institute of Electrical and Electronics Engineers, Inc. All rights reserved.

Published by John Wiley & Sons, Inc., Hoboken, New Jersey.
Published simultaneously in Canada.

No part of this publication may be reproduced, stored in a retrieval system, or transmitted in any form or by any means, electronic, mechanical, photocopying, recording, scanning, or otherwise, except as permitted under Section 107 or 108 of the 1976 United States Copyright Act, without either the prior written permission of the Publisher, or authorization through payment of the appropriate per-copy fee to the Copyright Clearance Center, Inc., 222 Rosewood Drive, Danvers, MA 01923, (978) 750-8400, fax (978) 750-4470, or on the web at www.copyright.com. Requests to the Publisher for permission should be addressed to the Permissions Department, John Wiley & Sons, Inc., 111 River Street, Hoboken, NJ 07030, (201) 748-6011, fax (201) 748-6008, or online at http://www.wiley.com/go/permission.

Limit of Liability/Disclaimer of Warranty: While the publisher and author have used their best efforts in preparing this book, they make no representations or warranties with respect to the accuracy or completeness of the contents of this book and specifically disclaim any implied warranties of merchantability or fitness for a particular purpose. No warranty may be created or extended by sales representatives or written sales materials. The advice and strategies contained herein may not be suitable for your situation. You should consult with a professional where appropriate. Neither the publisher nor author shall be liable for any loss of profit or any other commercial damages, including but not limited to special, incidental, consequential, or other damages.

For general information on our other products and services or for technical support, please contact our Customer Care Department within the United States at (800) 762-2974, outside the United States at (317) 572-3993 or fax (317) 572-4002.

Wiley also publishes its books in a variety of electronic formats. Some content that appears in print may not be available in electronic formats. For more information about Wiley products, visit our web site at www.wiley.com.

Library of Congress Cataloging-in-Publication Data

Names: Slavin, Lawrence M., author.
Title: Overhead distribution lines : design and application / Lawrence M. Slavin.
Description: Hoboken, NJ : Wiley, [2021] | Includes bibliographical references and index.
Identifiers: LCCN 2020031214 (print) | LCCN 2020031215 (ebook) | ISBN 9781119699132 (cloth) | ISBN 9781119699194 (adobe pdf) | ISBN 9781119699200 (epub)
Subjects: LCSH: Overhead electric lines–United States–Handbooks, manuals, etc. | Electric lines–Poles and towers–Handbooks, manuals, etc. | Telecommunication lines–United States–Handbooks, manuals, etc.
Classification: LCC TK3231 .S57 2021 (print) | LCC TK3231 (ebook) | DDC 621.319/22–dc23
LC record available at https://lccn.loc.gov/2020031214
LC ebook record available at https://lccn.loc.gov/2020031215

Cover design by Wiley
Cover image: © stevenfoley/Getty Images

Set in 9.5/12.5pt STIXTwoText by SPi Global, Chennai, India
10 9 8 7 6 5 4 3 2 1

I dedicate this book to two groups of people. First of all, to my wife, Helen, our children and their spouses, and, of course, our grandchildren, who are all quite special. I also dedicate this manual to the colleagues and friends that I have known and had the pleasure to work with over the past decades. I have had the good fortune of being able to continue to pursue my technical interests and activities into my "semi-retirement" years, while participating in professional organizations, as well as providing services on behalf of my clients, who have treated me with courtesy and respect.

Contents

About the Author   *xi*
Preface   *xiii*
Acknowledgments   *xv*

1   **Introduction**   *1*
1.1   Scope   *1*
1.2   Background   *2*

2   **Pole Structures**   *7*
2.1   General   *7*
2.2   Wood Pole Strength   *8*
2.3   Loads   *13*
2.4   Embedment Depth   *15*
2.5   Guying   *17*
2.6   Column Buckling   *19*
2.7   Grounding and Bonding   *22*

3   **Pole Installation and Maintenance**   *23*
3.1   Pole Placement   *23*
3.2   Guys and Anchors   *24*
3.3   Pole Maintenance   *26*

4   **Wires, Conductors, and Cables**   *31*
4.1   Categories   *31*
4.2   Messenger Wire/Strand   *31*
4.3   Electric Supply (Power) Cables   *33*
4.4   Communications Cables   *35*
4.5   Wireless Attachments   *38*

5   **Cable Installation**   *39*
5.1   Conductor and Cable Placement   *39*
5.2   Lashing Operation   *40*
5.3   Overlashing   *41*

6   **NESC® Requirements (Strength and Loading)**   *45*
6.1   National Electrical Safety Code (NESC)   *45*
6.2   Loading Requirements   *46*
6.3   Strength Requirements   *49*
6.4   Wire Tensions   *52*
6.5   Guyed Poles   *53*
6.6   Extreme Wind Loads ("60 ft Limit")   *54*
6.7   Allowable Deterioration   *56*
6.8   Overlashed Cables   *57*
6.9   Software Tools and Pole Loading Analysis   *60*

7   **NESC® Requirements (Clearances)**   *63*
7.1   Clearances   *63*
7.2   Clearance Zones   *63*
7.3   Clearances Above Surfaces and Buildings   *66*
7.4   Clearances Between Wires   *67*
7.5   Overlashed Cables   *67*

8   **Principles of Wire Sag**   *71*
8.1   Catenary   *71*
8.2   Initial and Final Sag   *72*
8.3   Sag–Tension Relationship   *72*
8.4   Determining Change in Sag (and Tension)   *73*
8.5   Ruling Span   *76*
8.6   Point Load   *77*

9   **General Order 95 (California)**   *81*
9.1   General Order 95 (GO 95)   *81*
9.2   Loading Requirements   *81*
9.3   Strength Requirements   *83*
9.4   Clearances   *84*

10   **Examples**   *85*
10.1   Purpose   *85*
10.2   Tangent Line   *85*
10.3   Line Angle   *88*

10.4   Line Angle – Buckling Consideration   *90*
10.5   Additional Attachment   *94*
10.6   Summary   *96*

Appendix A   Properties of Messenger Strands   *99*

Appendix B   Wireless Attachments   *101*

Appendix C   Extreme Wind and Extreme Ice Loadings   *103*

Appendix D   Solution of Cubic Equation   *107*

Appendix E   Point Load   *109*
E.1   Parabolic Model   *109*
E.2   Intersecting Straight Lines Model   *111*

Glossary   *115*
References   *121*
Index   *123*

About the Author

Dr. Lawrence (Larry) M. Slavin is Principal of Outside Plant Consulting Services, Inc. which was established in the year 2002 to help meet the needs of the telecommunications and power industries in establishing standards, guidelines, and practices for outside plant (outdoor) facilities and products. Larry has extensive experience and expertise in such activities, based upon his many years of service at AT&T/Lucent Bell Telephone Laboratories (Distinguished Member of Technical Staff) in telecommunications product design and development, followed by a career at Telcordia Technologies (now Ericsson, formerly Bellcore) in its research and professional service organizations. As Principal Consultant and Manager/Director of the Network Facilities, Components and Energy Group at Telcordia, he provided technical leadership in developing installation practices and "generic requirements" documents, introducing new construction methods, and performing analyses on a wide variety of technologies and products. Throughout his long career, Larry has had a leading role in the evolution of many telecommunications related fields and disciplines, for both overhead and belowground lines. Larry received his Bachelor's degree in mechanical engineering from The Cooper Union for the Advancement of Science and Art, and his Master's and PhD degrees from New York University. He lives with his wife, Helen, in White Meadow Lake in north central New Jersey.

Preface

This manual provides a general overview of the use of utility distribution poles for electric supply and communication applications. The intended audience includes utility engineers as well as those without a technical background, but who desire an understanding of the principles and various issues related to their application. The book is more an introduction to the subject than a detailed design, installation, or operations handbook and is not a replacement for the normal training and experiences typically required for either power or communications engineers responsible for the design, operation, and maintenance of overhead lines. This manual is a compendium of practical engineering and design information for poles and supported facilities, enhanced with some technical details that are not readily available elsewhere, that will prove to be helpful in some utility applications. The book is not intended to cover construction practices, which items are appropriately covered elsewhere, such as The Lineman's and Cableman's Handbook and the Telcordia Blue Book – Manual of Construction Procedures.

Information is provided on the physical characteristics of various types of utility poles, overhead supply and communication wires and cables, joint-use issues and related safety rules, including national standards and other related documents. The single most relevant document is the National Electrical Safety Code (NESC®), which is adopted or reflected in various regulations or industry practices throughout most of the United States and its territories. Other relevant documents include national standard ANSI O5.1 Wood Poles: Specifications and Dimensions and as well as various ASCE manuals, and documents produced by the Rural Utilities Service (RUS). The State of California is unique in that it has produced its own rules prescribed for overhead lines under the jurisdiction of its Public Utilities Commission, General Order No. 95 Rules for Overhead Electric Line Construction.

It was the original intention that this manual be published to include any new information contained in the upcoming (2022) edition of the NESC. However, the possible delayed availability of this edition, resulting from the COVID-19 pandemic, suggested that it would be more beneficial to base the manual on the current (2017) edition of the NESC, presented in a manner consistent with the publically available Pre-Print for the 2022 edition, rather than delay the publication of this manual. It is also recognized that, in most cases, based on the "grandfather" clause of the NESC, any new rules would only be applicable to newly constructed pole lines.

The information in this document is believed to be accurate at the time of its preparation, but is provided without any warranty, expressed or implied.

Lawrence M. Slavin, PhD
New York University, 1969

Acknowledgments

The author expresses his sincere appreciation to those that helped facilitate this publication, ranging from the courteous and efficient staff at Wiley-IEEE Press, and Ernesto Vega Janica at IEEE, to those that provided a technical review of this manual. In this regard, I particularly thank Trevor Bowmer and Dave Marne for their detailed formal review of the entire manual, which I recognize required a significant effort. I am also grateful to Martin Rollins, Andy Stewart, and Sam Stonerock for their review of specific portions of the manual. I consider myself fortunate to have colleagues that were willing to devote the extra time and energy, in addition to their many other obligations, to help me complete this project. However, any possible errors in the published document are my own responsibility, in spite of the best efforts of the reviewers.

1

Introduction

1.1 Scope

This manual presents a brief description of typical practices for the design, installation, and usage of overhead utility distribution lines, providing an understanding of the basic principles, and facilitating the subsequent pursuit of the technology and issues in greater detail, as desired or appropriate. It is not, however, the intention of this document to provide or duplicate existing detailed design, construction, or installation specifications and information as presently employed by the various electric supply and communications utilities for their overhead facilities. Such information is available within the individual utilities or their representative organizations.

Although some of the information in this manual is applicable to a variety of overhead applications, including high-voltage transmission lines, the focus of the information is on the local distribution systems, closer to the customer or subscriber end of the grid. This portion of the network is typically characterized by efficient joint-use applications in which electric supply and communications facilities are in reasonably close proximity, supported on the same physical structures, but with strict rules and guidelines to help ensure reliable operation and the safety of both the public and the utility workers. These systems are characterized by significant variability between the facilities supported on each pole, and frequent changes based on customer need, and are generally not subject to the same level of physical design detail as the electrical transmission lines. Indeed, it would be neither practical nor cost-effective to perform a detailed, sophisticated structural analysis on every pole in a distribution line. In contrast, long-distance high-voltage power transmission facilities, along the same transmission line, are relative invariable and unchanging, with relatively uniform spans, and require the efforts of experienced structural engineers, often supported by the use of sophisticated software tools, to cost-effectively design reliable, safe overhead facilities.

Overhead Distribution Lines: Design and Application, First Edition. Lawrence M. Slavin.
© 2021 The Institute of Electrical and Electronics Engineers, Inc.
Published 2021 by John Wiley & Sons, Inc.

This manual is divided into 10 sections or chapters. The background, including a description of the two general categories of construction methods relevant to outdoor utility lines, is provided in Chapter 1, including the overall characteristics and relative advantages and disadvantages represented by each category. Chapter 2 contains a description of pole structures and their physical characteristics, while Chapter 3 provides an overview of their installation, operation, and maintenance procedures. A description of the various types of suspended wires, conductors, and cables is contained in Chapter 4, and Chapter 5 discusses the methods of their installation. Chapters 6 (Strength and Loading) and 7 (Clearances) are primarily based on the rules of the National Electrical Safety Code (NESC®), which govern the physical design and construction of the overhead lines, to help ensure safe facilities for the public and utility workers. Chapter 8 explains the principles regarding wire sags (and related tensions) and provides methods for determining these values under various weather and operating conditions. Chapter 9 contains a brief description of General Order 95, which governs overhead utility lines in California, and how it compares to the NESC. Example calculations are included in Chapter 10 for estimating the physical status of pole systems subject to storm loads. Additional details and explanatory information supporting the various chapters is contained in Appendices A–E. A glossary of terms and a list of references are also provided.

1.2   Background

There are two basic modes of construction for outdoor utility lines for electric power supply and communication systems:

(1) Belowground (or underground) plant consisting of an array of parallel conduit paths, spanning the distance between manholes, typically located parallel to a main thoroughfare or highway; or, for more local distribution applications, direct burial of the cables within the soil, possibly placed along a road or street, with buried service drops to the residences. Routine access to such distribution facilities is typically provided by flush-mounted handholes or above ground terminals and pedestals.

(2) Aerial/overhead plant in which the cables are individually suspended between utility structures, including tall structures or towers for high-voltage transmission lines, spaced hundreds, or possibly thousands, of feet apart; or relatively short structures for distribution applications, typically individual poles, spaced up to a few hundred feet apart.

For both these modes of construction, the requirements and guidelines for the installation and operation of power supply and communications lines, including the shared usage of facilities (poles, underground conduit systems, trenches), are provided in utility industry standards and documents, the most significant of which is the NESC. Both construction methods are commonly used in the industry, with an increasing amount of belowground distribution facilities being placed relative to aerial plant in more recent decades, primarily driven by regulations. For example, the large majority of new construction in local (residential) subdivisions deploys belowground facilities, in response to the demands of communities and various levels of government for a greater portion of belowground construction along roads and thoroughfares, primarily due to esthetic considerations and safety concerns. Utility pole collisions account for a significant fraction of automotive fatalities along the nation's roads and highways, for which the U.S. Federal Highway Administration, as well as the individual states, therefore provide appropriate guidelines for their usage (FHA 1993; AASHTO 2011).

A description of belowground cable applications, including conduit and duct applications, is provided in ASCE Manual of Practice No. 118 (ASCE 2009). Figure 1.1 illustrates typical belowground (underground conduit and direct-buried) construction alternatives.

The extensive conduit facilities of underground plant are generally appropriate for limited applications, such as associated with the trunk or feeder portions

Cables installed
within ducts

Source: Courtesy of Underground Devices, Inc.   Source: Lawrence M. Slavin

Figure 1.1 Typical belowground construction.

of the traditional telecommunications network, owing to the high cost of this method of construction. However, the availability of several vacant conduits does provide flexibility, including the capability to postpone installation of expensive trunk or feeder facilities (fiber-optic, etc.) until the need arises. Such expensive underground conduit systems are also the only viable alternative in metropolitan or large urban areas where overhead lines and/or future digging are not practical options. In comparison, direct-buried plant is a lower cost method for placing individual cables belowground between any desired termination points, but lacks any flexibility with respect to future additions or replacements. While the placement of utility lines belowground, using either method of construction, avoids much of the potential damage resulting from extreme weather events, there is nonetheless greater vulnerability during incidents of flooding and accidental damage during excavation work in the area.

Although not esthetically pleasing, the ubiquitous overhead lines throughout the United States – supported by possibly as many as 200 million utility poles – provide many important benefits, and is the reason these structures and suspended lines continue to be widely used. Individual distribution poles, or even lattice transmission towers, require minimum real estate at the ground level, and allow new lines to be readily deployed in available overhead space. This includes otherwise difficult crossing applications, or where expensive belowground construction methods (e.g. directional drilling) would be required, such as at highways, railroads, and waterways. Overhead installations avoid the many possible issues encountered when attempting to perform construction beneath the surface in various or unknown belowground conditions, often in the presence of existing belowground facilities. The latter situation can be particularly hazardous, especially when power or gas lines are in the vicinity. The use of mandatory "call-before-dig" rules, and related utility locating practices, are not infallible, and unfortunate accidents may occur in spite of such precautions.

Apart from natural or man-made disasters, overhead lines are exposed to environmental stresses that are generally less severe than the persistent wet and corrosive surroundings that can be found belowground. As a result, it is often more of a challenge to design the belowground cables and/or the associated conduit/manhole facilities with sufficient resistance to those degradation forces than where the cables are placed overhead. In addition, if degradation occurs, or water penetrates the belowground plant, their repair and replacement is more difficult, expensive, and time-consuming. In general, overhead lines are inherently significantly less expensive to install and maintain than belowground facilities, as well as being characterized by greater flexibility for the addition, rearrangement, and/or replacement of the supported lines and equipment.

Figure 1.2 illustrates a typical distribution utility pole application including sharing, or joint-use, of the pole for supporting electric power supply and communications (telephone, Cable/CATV) lines.

Figure 1.2   Typical joint-use utility pole application. Source: Lawrence M. Slavin.

It is recognized that the increasing deployment of wireless technologies (cellular phones, satellite TV, etc.) has greatly impacted the communications industry, resulting in lost revenue for some of the wireline-based utilities. Nonetheless, it will be a very long time, if ever, before cable-based wireline communications become replaced and discarded. Wireless technology is inherently inferior to wireline (copper, fiber, coax) technologies with respect to various characteristics and features (security, reliability, quality, information capacity, etc.), thereby inhibiting elimination of physical cables. The present major investment in new wireline facilities by the major telephone companies, wherever feasible, bears witness to this principle. Furthermore, "wireless" systems contain multiple wireline segments, such as for interconnecting towers and cell sites, and for providing backhaul communications to the necessary central offices and data centers, as well as for supplying power to the wireless facilities. Ironically, the deployment of the latest wireless technology requires denser placement of antennas than previous systems, encouraging their installation at a larger number of elevated locations, for which utility structures and poles are prime candidates. In particular, the much vaunted 5G wireless age relies on these utility poles as the required fundamental infrastructure to provide the comprehensive coverage necessary for their new services, for which the physical issues discussed in this manual should be considered. In addition, although various alternatives are being pursued for distributed, renewable energy sources for electric power, the continued use of physical cables for transporting electric power supply to homes and industries will be required for the indefinite future.

2

Pole Structures

2.1 General

Distribution (and transmission) poles are available in various materials, with wood being the most commonly used, especially for distribution poles. Due to their prevalence, some of this document will inevitably include information pertaining to wood products, but should not be interpreted as an endorsement relative to other materials. Other pole materials that may be deemed appropriate, depending upon the specific application, include engineered products such as steel, concrete, and fiber-reinforced polymer. Each pole material has its particular advantages, relating to availability, cost, esthetics, installation, maintenance, longevity, etc.

Although the basic distribution pole is a simple, usually tapered, columnar structure, numerous auxiliary hardware items are necessary to accomplish a practical installation. Figure 2.1 shows many such items, such as

- Crossarms and support braces
- Ground rods
- Guys and anchors
- Various bolts, washers, hooks, etc.

in addition to a variety of brackets, clamps, etc. It is not the intention of this document to provide a detailed list of all types of supplier products that may be used as part of an overhead distribution system. Such information is readily available from utility product manufacturers and vendors. The poles themselves are described in some detail below, including their physical properties and application information, the most common of which are naturally grown wood poles.

Overhead Distribution Lines: Design and Application, First Edition. Lawrence M. Slavin.
© 2021 The Institute of Electrical and Electronics Engineers, Inc.
Published 2021 by John Wiley & Sons, Inc.

2.2   Wood Pole Strength

Specifications for naturally grown wood utility poles are provided in the ANSI O5.1 standard, including the widely used strength classification system that provides the average strength or capacity based upon a lateral (horizontal) load applied 2 ft from the tip of the pole, acting as a vertically mounted cantilever, as illustrated in Figure 2.2, independent of the length of the pole or its wood species (ANSI 2017).

Table 2.1 defines the "Class Loads." (The H-class poles are generally not required or used for distribution applications.) The class load assumes the peak stress in the pole will occur at the groundline (GL), for which the minimum GL circumference is determined accordingly, using the basic formula:

$$M_{GL} = \left(\text{fiber strength}\right) \, C_{GL}^3 / \left(12 \times 32 \, \pi^2\right)$$

where C_{GL} is the GL circumference (in), M_{GL} is the GL bending moment (ft-lbs), and the "fiber strength" (lbs/in^2) is provided in the ANSI O5.1 standard for each wood species (e.g. average 8000 psi for Southern Pine and Douglas Fir, and 6000 psi for Western Red Cedar). The minimum pole dimensions (circumferences) for each class, as specified 6 ft from the butt, are based on the pole length, required GL circumference, assumed taper and embedment depth (typically 10% of the length plus 2 ft; see Section 2.4). Longer (taller) poles of the same class have a larger GL diameter to withstand the correspondingly greater groundline bending moment. For example, a "4/40" pole designates a Class 4, 40 ft long pole, with an average load capability of 2400 lbs, and has a greater GL circumference than a "4/35" (Class 4, 35 ft long) pole.

In principle, due to the tapering of the pole, the peak stress may theoretically occur at a point somewhat above the groundline; i.e., for an ideal truncated conical prism, the peak stress will occur at a point where the local circumference is 1½ times that at the point of load application. Furthermore, there is a tendency for the fiber strength to decrease with height. The combination of these two effects could result in a pole strength (allowable cantilever load) less than the class load. However, commercially available utility poles for typical distribution applications – e.g., 55 ft and shorter – are generally oversized beyond their minimal required dimensions, and the point of peak stress occurs at or near the groundline, and/or the pole will support their class load, regardless of the rupture location (ANSI 2017). For taller poles, for which the cables are distributed along the length, including joint-use applications with communications cables mounted closer to the middle of the pole, in addition to the power cables located near the top of the structure, the peak stress will often still occur at or near the groundline.

The concept of the wood pole class and related discussion of location of peak stress are based upon lateral loads applied to the pole, acting as a cantilever, and therefore does not directly account for vertical loads. Except for guyed poles,

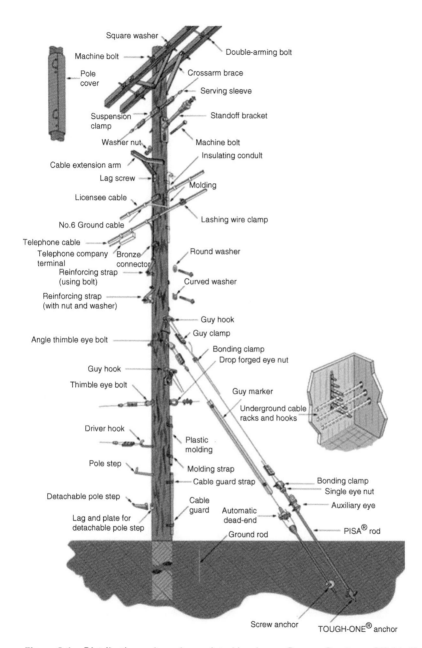

Figure 2.1 Distribution pole and associated hardware. Source: Courtesy of Hubbell Power Systems, Inc.

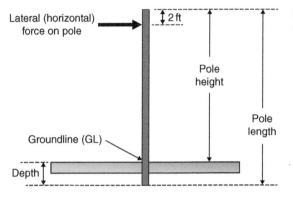

Lateral (horizontal) force on pole

2 ft

Pole height

Pole length

Groundline (GL)

Depth

Figure 2.2 Basic pole load geometry (class loads).

Table 2.1 Class loads for wood poles.

Class	Lateral load (lbs)
H6	11 400
H5	10 000
H4	8 700
H3	7 500
H2	6 400
H1	5 400
1	4 500
2	3 700
3	3 000
4	2 400
5	1 900
6	1 500
7	1 200
9	740
10	370

which guy tensions may generate extremely high vertical loads, possibly resulting in buckling (Section 2.6), and large eccentric vertical loads, the direct effect of the supported weights, as distributed over the pole cross-sectional area, is usually negligible compared to the associated bending stresses. Thus, the class load multiplied by the assumed height of the class load results in an allowable

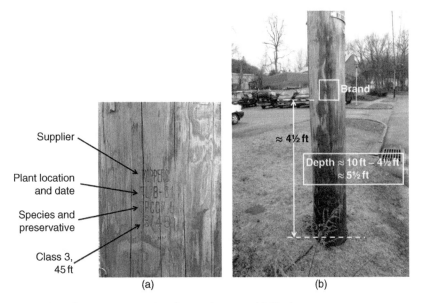

Figure 2.3 Brand/mark on poles. Source: Lawrence M. Slavin.

GL bending moment, providing a measure of the pole strength. This strength is also useful to compare to the effects of more general loads, which may include additional bending moments from large eccentric vertical loads which may have an appreciable effect, including possibly significant "*P*-delta" effects (Section 2.3), providing the peak stress from the cumulative bending moments occurs at or near the groundline. This may conveniently be assumed to be the case for poles 55 ft and shorter based on the combination of the oversize effect plus the effect of the communications cables mounted lower on the pole.

The pole class and size (length), as well as other product information (species, preservative, year of manufacture, etc.) is branded onto the surface of the pole, as illustrated in Figure 2.3.[1] The bottom of the brand or mark is located 10 ft above the butt for poles up to 50 ft in length, and 14 ft above the butt for longer poles. This location of the brand therefore allows a determination of the depth of embedment, as also indicated in Figure 2.3.

Since the wood classification system is widely understood and used, there is a tendency for the utilities to request the "equivalent" class of engineered poles being proposed as alternatives or replacements for conventional wood poles. The most direct comparison would presumably be based upon the same average strength of a similar length pole. However, this direct comparison is inhibited by

---

1 Figure 2.3 shows two different poles.

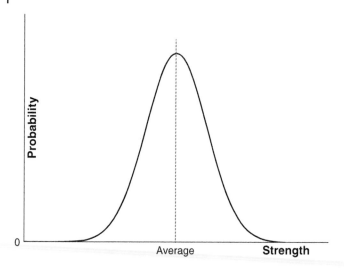

Figure 2.4 Wood pole strength variability.

the specification of a minimum[2] (vs. average) strength of an engineered product, and especially by the lower variability of engineered products, in comparison to the significantly greater variability (i.e. coefficient of variation, or relative standard deviation) of naturally grown wood poles; see Figure 2.4. It would therefore be reasonable to compare the designated (minimum) strength of the engineered (e.g. non-wood) pole to a somewhat lower (than average) strength of the given class wood pole, to achieve similar performance and reliability. Unfortunately, the quantitative determination of the lower designated strength is not obvious, and, in general, there is no universal equivalency between the products, and any desired "equivalency" can only be made in the context of an assumed loading condition and associated characteristics.

Previous attempts have been made to erroneously determine an "equivalency" by use of the strength ("derating") factors specified in the National Electrical Safety Code (see Section 6.3). Different strength factors are applied to wood material as a function of the Grade of Construction (Grade B or C), which relates to a desired level of reliability. In this comparison, the presumed (derated) strength of the same class of wood pole for a higher grade of construction (Grade B) is lower than that presumed for a lower grade of construction (Grade C), which may then be erroneously used to suggest that the (minimum) designated strength of the engineered poles can merely match the lower (Grade B) assumed (derated) strengths of the same class of wood poles that may otherwise be deployed for the commonly

---

2 The "minimum" strength is often considered to be the lower fifth percentile strength of the product.

used lower grade (Grade C) applications. This paradox arises from the incorrect assumption that the class (i.e. average strength) of the naturally grown wood pole is the important parameter as opposed to the strength required to support the specified load, and the subsequent confusion between the *presumed* (derated) strengths of widely variable wood poles with the *actual*, narrow range of strengths of the engineered poles. The relatively low derated strength of wood poles allows many such poles to withstand considerably greater applied loads than may be specified, or initially anticipated, in contrast to the more narrowly confined strengths of engineered products.

This issue of determining a technically correct equivalency has been addressed by using a reliability-based design (RBD) process, in which the statistical properties of the naturally grown wood and engineered poles are considered relative to the statistics of specific storm loads, resulting in similar failure rates, for two assumed levels of reliability (ASCE 2006). The results of this study indicated that a wood strength factor of 0.79, which is between those specified in the NESC for its two main grades of construction (i.e. 0.65 and 0.85 for Grade B and Grade C, respectively), would provide approximately equal levels of reliability between the wood and engineered materials, based on the particular storm loads (extreme wind, and extreme ice with concurrent wind) and loading configuration (tangent structures) considered. In general, the appropriate equivalency (strength factor) would be different if based on other load configurations (e.g. axial buckling; Section 2.6).

2.3 Loads

The lateral or transverse loads experienced by a pole include those resulting from wind loads on the wires or conductors, which loads are transferred to the pole, and wire tensions at line angles (or dead-ends), as well as the wind loads directly applied to the pole, which are relatively low compared to those due to the conductors and wires. Figure 2.5 illustrates these loads, as well as the vertical (weight) loads on the wires, which loads are also transferred to the pole, plus the weight of supported equipment. For "tangent" structures (little or no line angle), the wire tensions are generally balanced in opposite longitudinal directions, with no net load. Thus, the critical loads are usually the transverse loads caused by wind. For structures at significant line angles, the critical loads also include the transverse components of wire tensions. (Dead-ends are an exception, for which the unbalanced longitudinal wire tensions are a major consideration.)

In principle, the loads resulting from the wind and weight loads on the wires are transferred to the pole via the wire tensions, which is not apparent in Figure 2.5, since the tension vectors are conveniently shown acting along the chords spanning adjacent poles. In reality, the wire tension varies slightly in direction (and

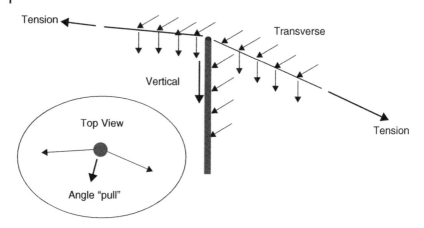

Figure 2.5   Pole loadings.

magnitude) along the span, by which the wind and weight loads are imposed on the structure.

Generally, except for the possible impact on wire tensions, at line angles (or dead ends), the effects of vertical loads corresponding to the weight of conductors or equipment are relatively low compared to that the bending (flexural) effects of the transverse loads. However, the combination of a significant transverse deflection and a large overhanging weight – i.e., *P*-delta effect – may result in an appreciable increase in the effective bending loads and stresses. Some design procedures account for this effect, such as where heavy facilities (e.g. transformers, or long spans of multiple ice-covered conductors) are mounted near the top of the pole, possibly aggravated by an eccentric mounting. The contribution of communications lines to this effect is typically small because of their low eccentricity when mounted on the surface of the pole, and lower attachment heights on joint-use poles.

Figure 2.6 illustrates the *P*-delta effect, including an equation for the total groundline bending moment, for which the product of the "weight" (designated by "*P*") and "lateral deflection" (designated by the Greek symbol "Δ," or "delta") represents the additional contribution to the bending moment. Since the deflection is much smaller than the applied height(s) of the transverse forces, the second term ("*P*-delta") in the equation for calculating the bending moment is often ignored in comparison to the first term, especially for distribution applications. Analyses that consider the *P*-delta contributions are considered to be "nonlinear" since the stresses in the pole are more complicated than may be represented by a simple "linear" sum of those corresponding to the individual applied loads.

Chapter 6 discusses the mandated load and strength requirements as provided in the NESC.

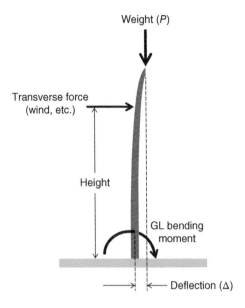

Weight (*P*)

Transverse force
(wind, etc.)

Height

GL bending
moment

Deflection (Δ)

GL bending moment =
Transverse force × Height + Weight (**P**) × Lateral deflection (**Δ**)

"**P – Δ**"

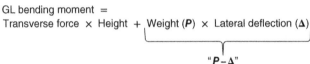

Figure 2.6 Bending moment, including "*P*-delta" due to lateral deflection.

2.4 Embedment Depth

The traditional industry rule-of-thumb is that the embedment depth for a wood pole is 10% of the length plus 2 ft. While these distances are indicated in the ANSI O5.1 standard, recent editions emphasize that these distances are not necessarily embedment depths, but are mainly provided to allow definition of other specifications for the poles – including the determination of class loads, as illustrated in Figure 2.2.

In addition to embedment depth, the stability of the pole, regarding overturning, clearly depends on other parameters, most obviously the type of soil. Based on extensive field experiments, the Rural Utilities Service (RUS) provides the following formula for determining appropriate embedment depths for wood transmission poles (RUS 2015):

$$P = S_e D_e^{3.75} / \left[L - 2 - 0.662\, D_e \right]$$

where P is the lateral force applied 2 ft from the top, that will overturn the pole, L is the total pole length (ft), D_e is the embedment depth (ft), and S_e is the soil constant. The soil constant S_e is equal to 140 for good soils, 70 for average soils, and 35 for poor soils, where such "soil" types are described in the RUS document based on the ability to provide firm support. The RUS recognizes, however, that local experience should be considered, which is important since detailed knowledge of the soil conditions may not be readily available, especially for typical distribution applications. Furthermore, since soil is not a controlled engineering material, it would not be reasonable to solely rely on any related theoretical models.

An interesting aspect of this model is that the overturning force, or applied GL moment (Figure 6.3), as determined by the formula, is not dependent upon the diameter of the pole, which relates to the belowground bearing surface. This phenomenon arises from the observation that the overturning is resisted by a large clump (prism) of soil, the size of which is relatively insensitive to the pole diameter (Seiler 1932), at least for the wood pole applications.

The overturning force calculated by the formula may then be compared to the class load, or breaking strength, which is also related to the same GL (bending) moment. Based on these results, and depending upon the soil conditions, pole length, and class size, the traditional rule-of-thumb may not be sufficient to allow the pole to develop its full strength without overturning (RUS 2015). This is as anticipated since the strength of the poles of a given length will vary with the class size, corresponding to increasing strength (i.e. maximum GL bending moment) at the larger sizes, which may then fail by overturning prior to breaking (rupture), especially in poor soils.

Although the RUS provides the above formula for wood pole applications, it proposes a different methodology for concrete and steel transmission poles (RUS 2017). The considered sizes (diameters) of these non-wood poles are in the range of 1–4 ft, much larger than that for wood poles, and the results are therefore presented as a function of the diameter. A comparison between the results, considering the non-wood poles at the low end of the range, such as would be applicable for distribution applications, indicates that the recommended depths for the model intended for the concrete and steel poles are significantly more conservative (greater) than that provided by the wood pole model described above, although both are based upon stability (overturning) considerations. The quantitative differences between the wood and non-wood models may be related to the observed clump of soil associated with the overturning of the wood poles. Nonetheless, this inconsistency, in addition to potential lack of reliable information on soil conditions, emphasizes the importance of local experiences in selecting an appropriate embedment depth for typical distribution applications (see **Appendix B**). In spite of these apparent discrepancies between the quantitative results of the two methods, both models reflect the low effect of the GL shear force compared to the GL moment in the tendency to overturn the pole.

2.5 Guying

Guying is required when the physical strength of the pole is not sufficient to support the loads that may be applied to the pole, or when it may be desired to limit deflections. These loads may be everyday forces that are not balanced, such as at a line angle (or dead-end), or possibly due to a heavy asymmetrically mounted transformer, which forces would be aggravated under storm load conditions. Guys may be installed perpendicular to the general direction of the line, such as at line angles, or parallel to the line, such as at dead-ends, or possibly to prevent cascading in the longitudinal direction in the event of a failure of conductors. Guys may also be installed, at various angles to the direction of the line, as deemed necessary, including to provide additional support to withstand storm loads.

Ideally, the guys should be attached at the same height of the pole as the conductors or wires (e.g. messengers) that are applying their unbalanced load, such as indicated in Figure 2.7 for the dead-end application, since this would directly resist the unbalanced load and avoid local bending effects on the pole. In practice, however, this is not always feasible, especially when there are multiple wire attachments, such as in the increasingly busy communications portion of the pole, located below the supply lines (see Figure 7.1). If less guys are used, each guy should be attached toward the center of the group of conductors or wires it is intended to balance. Separate guys should be used for the electric supply and communications portions of the pole, although the same guy anchor(s) may be used for both types of guys, depending upon the anchor capacity.

Guys are applied at an angle to the vertical structure, as defined by the "lead" and "height" dimensions, as illustrated in Figure 2.7 for a simple dead-end configuration. The lower the lead/height ratio, the less efficient is the guy support, requiring

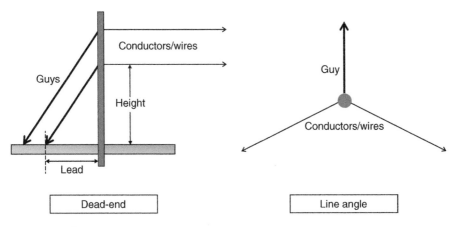

Figure 2.7 Guying.

increasingly greater tension to offset the tension(s) induced in the conductors or wires. Thus, a messenger strand (see Section 4.2) supporting a communications conductor may require a guy strand, and corresponding anchor, whose strength or capacity is several times that of the messenger strand itself, especially for relatively flexible structures, such as wood poles, or those of fiber-reinforced polymer, for which the guys are assumed to counteract essentially all of the intended load. For such cases, the following formula relates the guy (and anchor) tension for the dead-end, T_g, to that in the messenger, T_m, which may be considered to be either the everyday tension, or the peak (storm) tension (Section 6.4), depending on the load of interest:

$$T_g = T_m \frac{\sqrt{1 + R^2}}{R}$$

where R is the lead/height ratio (e.g. 1/5, 1/4, 1/3, 2/5, etc.), and is significantly less than 1 (unity) for most anchor guys. For values of R equal to 1/3 or lower, the guy tension is approximately equal to T_m/R, which represents an amplification of the messenger tension by the reciprocal of the lead/height ratio. This is a major effect which may impose a load of triple or greater the tension in the messenger on the required guy(s), as well as the anchor, which must be proportionally stronger than the messenger strand itself.

Figure 2.8 illustrates an unbalanced load resulting from wire tensions acting on a corner pole as in the line angle shown in Figure 2.7. In this case, the unbalanced load acting transverse to the pole, because of the (assumed) equal tensions on both adjacent spans, is given by the following formula:

Unbalanced transverse load (lbs) = Wire tension (lbs) × Pull (ft) / 50 ft

where the "pull" (indicated in Figure 2.8) is defined as the perpendicular distance along the angle bisector to the chord joining the points 100 ft along the spans on both sides of the corner pole. In this case, the guy tension T_g will again be amplified based on the lead/height ratio, as given by

$$T_g = T_{unbal} \frac{\sqrt{1 + R^2}}{R}$$

where T_{unbal} is the unbalanced load given above, as due to everyday tension or peak (storm) tension, again depending on the load of interest, and also includes the horizontal (transverse) wind load on the wires, as illustrated in the example in Section 10.3. The effect of the load amplification on the guys is reflected in the recommended guy strand sizes provided in the Telcordia *Blue Book – Manual of Construction Procedures* (Telcordia 2017) as a function of the messenger strand size and the line angle (pull). These guy sizes are quite conservative, consistent with the maximum possible tension (breaking strength) in the messenger strands,

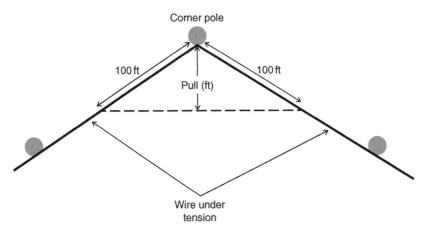

Corner pole

100 ft

100 ft

Pull (ft)

Wire under
tension

Figure 2.8 Unbalanced load on pole resulting from wire tension.

especially in consideration of the relevant strength and load requirements of the guy system; see Section 6.4.

Furthermore, the resulting guy strand tension will impose a significantly greater vertical (axial) load on the distribution pole than that typically caused by the weights of the supported conductors, cables, and equipment, including under ice-loading conditions. These high vertical/axial loads may precipitate buckling of the pole, acting as a column, as discussed in Section 2.6. The vertical component of the guy tension, T_v is given by

$$T_v = T_g / \sqrt{1 + R^2}$$

2.6 Column Buckling

Figure 2.9 illustrates three possible modes of column instability (i.e. buckling), which are precipitated by significantly different magnitudes of loads that may cause a failure. The theoretical (vertical or axial) buckling load W_{bu} is expressed by the general formula

$$W_{bu} = n\pi^2 E \, I_{eff} / h_{eff}^2$$

where E is the modulus of elasticity of the pole material, I_{eff} is the effective cross-sectional moment of inertia, h_{eff} is the effective height or length of the column, and the coefficient n is dependent upon the end support conditions. (The use of the appropriate units is illustrated in the Example in Section 10.4.) In particular, the value of the coefficient n can vary widely. For example, in Figure 2.9a, n equals $1/4$ where the upper end is unrestrained, in comparison to

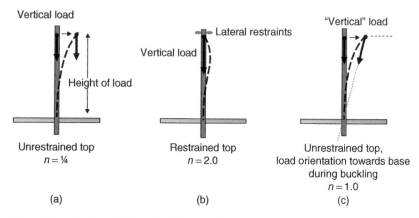

Figure 2.9   Modes of column buckling (instability).

a value of approximately 2.0 – almost an order of magnitude greater – with the upper end effectively restrained (i.e. "pinned"), as shown in Figure 2.9b, even while assuming the lower end is rigidly fixed at the base in both cases. The value of n is equal to an intermediate value of unity (1.0) when the axial load is directed toward the base of the pole while deflecting, as shown in Figure 2.9c. (A value of $n = 1$ also applies to a pole that is pinned at both ends.) The possible variability, or uncertainty, of the buckling load for utility pole applications is even greater when considering the distributed nature of the applied loads along the length of the pole and the nonconstant cross-section of a tapered pole, as well as the effect of deviations from the idealized models illustrated in Figure 2.9, such as possible non-rigid support at the groundline and the presence of even a small unbalanced lateral load (e.g. due to wind on the wires and pole). In general, engineering judgment is required to determine appropriate values of the various parameters for practical applications, including the effective heights and cross-sections, as well as the value of n, considering the line geometry and the corresponding guying configuration.

For the present purposes, it may be assumed that the guy wire(s) acting in combination with opposing tensions from the conductors or cables at line angles or dead-ends, effectively provides some restraint in movement at the top of the pole, in possibly both the lateral and longitudinal directions, depending upon the configuration. For this reason, an effective value of n of 1.5 has been historically implemented in guidelines for telecommunications applications for wood poles (ANSI 1995). This value appears optimistically close to the fully restrained case of Figure 2.9b, but the guidelines include other conservative, offsetting assumptions (effective column length, location of effective cross-section, etc.).[3] *RUS Bulletin 1724E-200* for transmission lines and *IEEE P751/D2 (Draft) Design Guide for Wood Power Transmission Structures* provides assumptions and methods for tapered poles, and recommends a value of *n* equal to 1.0 (RUS 2015; IEEE 1999), as a compromise value, for at least some guyed applications. These documents also provide guidelines for determining an effective value of the moment of inertia, I_{eff}, for solid tapered poles. The basic method ("Gere and Carter") corresponds to using an effective length equal to the height at the lowest load point, and adjusting the local moment of inertia by a factor such as $(D_{GL}/D_w)^2$, where D_{GL} is the diameter at the ground-line and D_w is the pole diameter at that load point. (The moment of inertia of a circular cross-section is proportional to the diameter raised to the fourth power; see Section 10.4.) The loads applied at higher points along the pole are directly added to that at the lowest load point.

Considering the significant differences in these methods, it may be expected that there would be major differences in the quantitative estimates of the buckling load and predictions of its occurrence. Sample calculations, however, do not indicate gross differences in the estimates and predictions, although the degree of discrepancies somewhat depends upon the location (attachment heights) of the guys and the number of guys. For cases with multiple guys at different heights, such as for communications and supply lines, the RUS method tends to be less conservative than the traditional telephone method, but both methods are believed to be conservative in such cases because of the additional restraints provided by the guys at multiple levels of the pole, which are not necessarily considered in the models. For the present purposes, the RUS model may be conveniently used, as illustrated in the example in Section 10.4.

It should be noted that for some load conditions, including possible test configurations for determining the buckling load of a pole, the effective value of *n* may depart from the cases discussed above because of the precise orientation of the load. For example, referring to the top unrestrained case of Figure 2.9c, if the actual "vertical" load is not the result of equipment or conductor weight, but corresponding to an applied strand tension aligned between the pole top and base (groundline support), causing a direct compression of the pole, the load may re-orient itself as indicated, tending to limit the lateral deflection. In this case, the value of *n* is equal to 1.0 (Timoshenko 1961), in contrast to the ¼ value otherwise applicable

---

3 The telecommunications industry has historically assumed the effective height of the applied load is measured from the butt, and that the critical section (for a tapered pole) is located 2/3 the distance from the butt to the (uppermost) load application point (ANSI 1995). For applications with several load attachment heights, the individual loads are converted to effective loads applied at the uppermost attachment point by applying an adjustment factor equal to the square of the ratio of the load distance (from the butt) to the maximum load application distance.

to the unrestrained case, illustrating the sensitivity of the results to the physical configuration.

An alternative approach would be to use software tools that are able to model the pole and calculate its deformation, including nonlinear effects. Such analyses reflect the additional effects of the loads acting on the deformed structure, providing an indication of the potential instability ("buckling") of the pole; see Section 6.9.

2.7   Grounding and Bonding

Proper electrical grounding and bonding of supply and communication lines is essential to help protect the public and utility workers. Appropriate requirements are contained in the National Electrical Safety Code (NESC 2017; IEEE 2019). Key points include:

- Grounding the neutral conductor of a multigrounded supply system and communications messengers at a minimum of four connections per mile, with some possible exceptions, not including the grounding at customers' service equipment;
- Supply and communications systems on joint-use poles should be grounded at the same poles, using a single grounding conductor, or to separate grounding conductors that are bonded together;
- Anchor guys must be grounded and/or insulated to help prevent the possibility of broken or slack guys becoming energized and become a hazard to the public, or communications workers on joint-use poles.

Additional details and explanations regarding the grounding requirements are provided in the *National Electrical Safety Code (NESC®) 2017 Handbook* (Marne 2017) and the *2017 NESC Handbook* (IEEE 2017). Related construction methods for accomplishing the grounding are provided in *The Lineman's and Cableman's Handbook* (Shoemaker 2017) and the Telcordia *Blue Book* (Telcordia 2017).

3

Pole Installation and Maintenance

3.1 Pole Placement

The poles themselves are typically and most conveniently installed using a pole derrick, which bores a vertical hole for inserting the structure, and then facilitates its placement (Figure 3.1). Section 2.4 discusses appropriate burial depths, depending on the pole and soil characteristics.

To the extent practical, as many accessories as possible, including crossarms and other hardware, should be attached to the pole prior to the erection process. Once the poles are installed, any additional required auxiliary items, such as illustrated in Figure 2.1, are mounted, in preparation for the stringing of the wires, conductors, or cables intended for the initial installation. Subsequent lines may require additional such hardware to be placed at that time.

In some cases, such as where access is difficult (e.g. rear yards) and/or a pole derrick cannot be used, a lightweight pole (e.g. fiber-reinforced polymer) may be advantageously used; see Figure 3.2.

Although poles would normally be installed to be as plumb (vertical) as possible, corner poles are initially placed with a deliberate slight angle (raked) to the vertical, in a direction opposite to the pull (Figure 2.8). For a pole that is intended to be guyed, the degree of offset at the top should be equal to the pole top diameter (slightly greater than $1/2$ ft). In this case, the guy strand is tensioned such that the subsequent tensioning of the suspension strand, including the effects of the cable placement, results in the pole stabilized in the vertical position. The tension in the guy strand may be adjusted as necessary, following the addition of future attachments. In some cases, such as at relatively small values of "pull", the pole may not be required to be guyed, possibly based on the use of stronger poles and/or deeper embedment. The pole would again initially be raked, but with the base placed approximately one foot in the direction of the pull and the top inclined such that subsequent tensioning of the suspension strand, including the effects of

Overhead Distribution Lines: Design and Application, First Edition. Lawrence M. Slavin.
© 2021 The Institute of Electrical and Electronics Engineers, Inc.
Published 2021 by John Wiley & Sons, Inc.

Figure 3.1 Hole boring and pole placement. Source: Courtesy of Altec Inc.

Figure 3.2 Installation of lightweight pole. Source: Lawrence M. Slavin.

the cable placement, results in the top of the pole stabilized at the desired corner (intersection) of the opposite sides of the pole line (AT&T 1968; Lucent 1996).

Following initial pole placement, the base is backfilled and thoroughly tamped in layers. Sufficient fill should be used to allow the water to drain away from the base of the pole. Where the pole is particularly vulnerable to damage from vehicles, such as parking lots, driveways, or alleys, the pole must be protected by either a guard placed on the pole itself or via barriers that isolate the pole from contact. Poles should also, as practical, not be placed in areas susceptible to fire, including from brush, grass, rubbish, or possibly from buildings.

For additional details regarding wood pole handling and setting, see *The Lineman's and Cableman's Handbook* (Shoemaker 2017).

3.2 Guys and Anchors

The need and use of guying is discussed in Section 2.5. The guys are attached at the mid and/or upper portions of the pole, as required to counteract the loads, using guy clamps or possibly just a wire wrap (Figure 3.3), connected to guy hooks mounted directly on the pole surface.

Source: Courtesy of Allied Bolt Products, LLC.

Source: Lawrence M. Slavin

Figure 3.3   Guy hook and sample application.

Figure 3.4   Guys and markers.
Source: Lawrence M. Slavin.

The opposite (lower) end of the guys are attached to guy/anchor rods which are connected to anchors embedded in the ground, as shown in Figure 3.4. Several guys (1, 2, or 3) may be attached to the same anchor rod. A guy marker is required for guys exposed to pedestrian traffic.

The guy wires themselves are generally of the same type and construction as the messenger wire described in Section 4.2, but – based on the discussion in Section

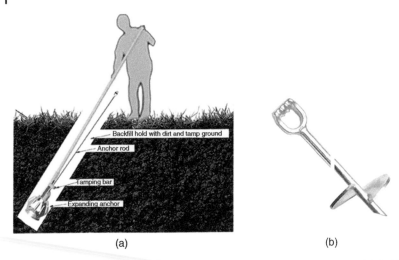

(a)   (b)

Figure 3.5 (a) Expansion anchor and (b) screw anchor. Source: Courtesy of Allied Bolt Products, LLC.

2.5 – may require sizes significantly larger than that of the messengers on the pole. The holding strength of the anchors must be consistent with the anticipated tensions induced in the guy wires. The NESC provides related strength requirements for guys and anchors.

There are several types of ground anchors, depending upon the application and soil conditions, and are available in a range of sizes and capacities, on the order of 10 000 lbs or greater. Figure 3.5 shows two of the widely used types of anchors for soil conditions. Anchors are also available for rocky or swampy conditions (Telcordia 2017).

3.3 Pole Maintenance

The practical life span of a utility pole is a function of the pole material and application environment, as well as the inspection and maintenance program. Thus, whereas some pole materials may appear to inherently have more longevity than naturally grown wood poles, long life spans for wood poles may be achieved by a combination of an initial preservative treatment, as appropriate, and an effective inspection and maintenance program. Indeed, it is not unusual for life spans of as much as 50 years, with an average of 35–45 years a reasonable expectation (Telcordia 2017). The ANSI O5.1 standard provides requirements for naturally grown wood utility poles, prior to the use of preservatives. Standards for the subsequent treatment by preservatives are provided by the American Wood Protection

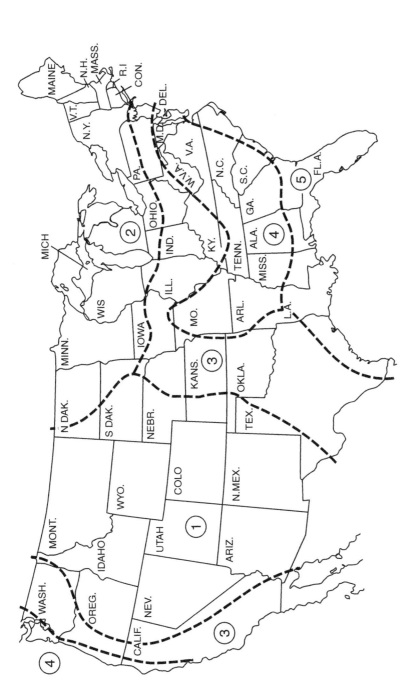

Figure 3.6 Decay severity zones for wood utility poles. Source: RUS Bulletin 1730B-121, provided by the USDA Rural Utilities Service.

Figure 3.7 Pole maintenance. Source: Courtesy of Osmose Utilities Services, Inc.

Association (AWPA 2019), as well by telecommunications industry requirements, including GR-60 (Telcordia 2011).

Inspections may be performed during routine field operations, or as part of a more formal inspection program. Such inspections may range from visual to sounding, to bore or prod, to excavation, depending upon the program and schedule established by the utility, and should be consistent with the age of the poles, and the likelihood of decay, although 10-year inspection cycles are not uncommon. Figure 3.6, provided by the USDA Rural Utilities Service, indicates the degree of vulnerability to decay within the 48 states. Zone 1 is the least vulnerable, while Zone 5 is the most vulnerable.

Relatively new poles (e.g. less than 5–15 years) would typically not require more than visual or sounding inspections, including those performed during normal operations. The suggested age range can vary based on deterioration zone, pole species, and original treatment. Since most deterioration is internal and/or external below the groundline, visual inspections are the least reliable. Sounding – i.e. by hitting the pole with a hammer – can provide more useful information, especially when performed by an experienced person. Boring is capable of directly determining internal decay, and may be performed at, or possibly below, the groundline, where deterioration is most likely to occur. Prodding with a pointed object at or below groundline is capable of identifying external decay. The most reliable method includes excavating below the groundline, where (external) decay may be observed and remedial methods, including the application of preservatives or other procedures, may most readily be employed (Telcordia 2011; RUS 2013; CPUC 2017). Figure 3.7 illustrates the use of excavation to inspect a pole and extend its useful life. While these procedures are generally intended to apply to wood poles, some variation may also be applicable to non-wood poles (e.g. visual, excavation, etc.)

In addition to the traditional methods of inspection described above, there are innovative techniques that attempt to quantify the status of the pole, such as those based on the gross vibration characteristics of the pole, or based on physical resistance to a small boring tool as relates to wood density. These methods are less invasive than excavating, but their usefulness requires further investigation and verification (see also Section 6.7).

(a)   (b)

Figure 3.8   Inspected poles. (a) Pole with inspection tag and (b) rejected pole. Source: Lawrence M. Slavin.

(a)   (b)   (c)

Figure 3.9 Rehabilitated or reinforced poles. Source: Lawrence M. Slavin.

Figure 3.8 shows the results of inspections of two different wood poles – one that passed inspection, possibly following some remediation (e.g. addition of supplementary preservative), and one considered beyond remedial action and to be replaced. Figure 3.9 shows examples of poles that have been rehabilitated or reinforced, allowing continued usage.

For additional details regarding wood pole inspection, assessment, maintenance, and repair, see ASCE Manual of Practice No. 141 (ASCE 2019).

4

Wires, Conductors, and Cables

4.1 Categories

Although their ubiquitous nature has led to utility poles being used to support additional equipment (luminaries, antennas, etc.), the basic purpose of the utility poles, and auxiliary hardware and equipment, is to support the conductors, cables, and wires spanning the distance between adjacent poles. In general, there are three types of wirelines:

- Messengers/strands
- Electric supply (power) conductors
- Communication cables

4.2 Messenger Wire/Strand

The size and type of messenger wire deployed depends on the application, including the span length and the conductor or cable sizes and weights to be supported. It will also depend on the design criteria, including anticipated loadings or otherwise required strength, as discussed in Chapter 6. Figure 4.1 shows a steel messenger wire for communication applications, which generally consists of seven individual smaller strands, the size of which determines the product breaking strength. The messenger wire is galvanized for environmental protection, which is at a higher level for corrosive areas, such as coastal regions or industrial areas. The same wire may also be used for guying applications.

The most common size of messenger or guy wire for communications applications is 6M (5/16 in.) or 10M (3/8 in.), although both larger and smaller sizes are available, ranging from 2.2M (3/16 in.), for supporting drop cables, to 16M (7/16 in.), as well as the rarely used 25M (1/2 in.). The "M" designation represents roughly 1000 lbs. Thus, the 6M wire has a rated breaking strength of 6000 lbs,

Overhead Distribution Lines: Design and Application, First Edition. Lawrence M. Slavin.
© 2021 The Institute of Electrical and Electronics Engineers, Inc.
Published 2021 by John Wiley & Sons, Inc.

Figure 4.1 Steel messenger wire. Source: Courtesy of American Wire Group.

Messenger plus hoops

Figure 4.2 Use of hoops to support heavy communication cable. Source: Lawrence M. Slavin.

whereas the 10M wire is rated at 11 000 lbs and the 16M at 16 000 lbs. Another size that has been widely used for communications applications is the relatively small (¼ in.) inexpensive 6.6M strand, composed of extra-high strength-steel wire, and rated at 6600 lbs (Telcordia 2017). Communications cables are typically lashed to the messenger with stainless steel wire, as described in Chapter 5, although hoops or rings have also been used in some cases, especially for relatively heavy, multi-pair copper cables (Figure 4.2). **Appendix A** provides the physical characteristics of typical messenger wires.

Electric supply cables may also be lashed to messengers, similar to communications applications, but other configurations are more commonly used. For example, messengers are used as an integral member of multiplex cables (Figure 4.4) or to support the individual conductors of the Hendrix Aerial Cable System described in Section 4.3. The messenger for multiplex cables is constructed of hard-drawn copper and is available in various sizes (gauges) and strength, ranging from 10 gauge for a duplex cable to as much as 4/0 gauge for a quadruplex configuration (Southwire 2009). The messenger also typically serves as a neutral.

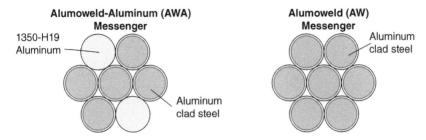

Figure 4.3 Electric supply messenger. Source: Courtesy of Marmon Utility LLC.

For some power applications, such as the Hendrix-type cable, the AWA messenger comprises a combination of individual aluminum strands and Alumoweld® (aluminum clad steel); see Figure 4.3a. This combination provides a balance between greater electrical conductivity and reasonable high tensile strength, as desirable for the system neutral in the Hendrix-type cable applications. Constructions consisting of only Alumoweld (AW) strands are also used where greater messenger strength is required, as illustrated in Figure 4.3b. These messengers are available in several sizes, ranging from 0.385 to 0.546 in. There is also a larger (0.722 in.), stronger AWA version with more strands of each type (12 Alumoweld, 5 aluminum) (Marmon 2018).

Wire constructions consisting of seven Alumoweld strands are also available for guy applications, in various sizes, including strength ratings equivalent to 6M, 10M, 16M, and 25M. The aluminum cladding provides protection against rusting, resulting in greater reliability and longevity and may be used for power or communications applications (AFL 2003).

4.3 Electric Supply (Power) Cables

Multiplex cables are commonly used for electric supply service drops and are available in various configurations, including duplex, triplex, and quadruplex. Figure 4.4 shows a quadruplex cable.

Individual covered conductors of the Hendrix-type supply cable, for providing primary distribution power, are shown in Figure 4.5. Assembled "cable" systems are shown in Figure 4.6, including the spacer/brackets for supporting the individual conductors on the messengers. The messenger typically also serves as the system neutral.

Most electric supply conductors and cables employed for both transmission and primary distribution applications are self-supporting and are available in a variety of types and constructions. Supply conductors may consist of solid copper or

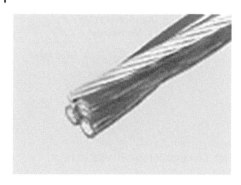

Figure 4.4 Multiplex (quadruplex) cable. Source: Courtesy of American Wire Group.

Figure 4.5 Hendrix covered conductors. Source: Courtesy of Marmon Utility LLC.

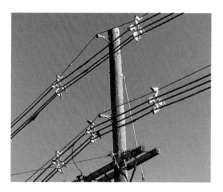

Figure 4.6 Hendrix covered conductors. Source: Lawrence M. Slavin.

aluminum wires, or stranded copper conductors, or special stranded cable construction, including (Southwire 2007):

- All-Aluminum Conductor (AAC)
- All Aluminum Alloy Concentric (AAAC) Lay-Stranded
- Aluminum Alloy Conductor Steel Reinforced (AACSR)
- Aluminum Conductor Aluminum Alloy Reinforced (ACAR)
- Aluminum Conductor Steel Reinforced (ACSR)
- Aluminum Conductor Steel Supported (ACSS)
- Aluminum Conductor Steel Reinforced/Trapezoidal-Shaped Wires (ACSR/TW)

Figure 4.7 shows examples of AAAC and ACSR/TW cables, which may be used for distribution or transmission applications.

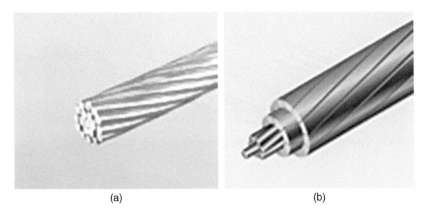

(a)   (b)

Figure 4.7 Electric supply conductors. (a) AAAC and (b) ACSR/TW. Source: Courtesy of American Wire Group.

4.4 Communications Cables

Communication technology is continuously advancing, for which the wireline transmission media has evolved from individual metallic wires to multipair cable to coaxial (concentric) conductors to optical fiber. Furthermore, the capabilities of the media are also continually improved as electronics allow more efficient usage of the media to carry greater information and deliver more bandwidth, as required for present and future applications. In spite of the dramatically increasing use of wireless signals, physical cables are still required to transport the collected and transmitted information from the local antennas to head ends, where they are forwarded to their destinations by other physical or wireless media. Indeed, the greater use of wireless systems, such as 5G, requires more equipment and cables to be mounted on distribution poles, which are becoming increasingly crowded.

Consistent with the variety of transmission media, service drops may comprise metallic pairs, coaxial cable or fiber. Figure 4.8 shows various types of self-supporting metallic service drops. Figure 4.9 illustrates a coaxial service drop and a connectorized fiber drop.

The distribution cables are generally larger in size, containing more copper pairs or fibers, or a larger diameter coaxial cable allowing greater bandwidth. Typical construction methods require the use of a separate steel messenger, as described above, but self-supporting designs with an integrated messenger, in a Figure 4.10 configuration, are sometimes used.

Conventional copper cables have been available in sizes up to hundreds, and even thousands, of twisted pairs, depending upon the wire gauge, and as much as several inches in diameter and several pounds per foot. Large numbers of such

Figure 4.8 Typical metallic service drops. Source: Courtesy of Superior Essex.

Figure 4.9 Typical coaxial and fiber service drops. Source: Courtesy of CommScope®.

Figure 4.10 Typical copper pair cables. Source: Courtesy of Superior Essex.

wires had originally been required when it was necessary to assign an individual pair to each telephone line. More recent digital transmission technologies have allowed multiplexing (combining) of lines on a reduced number or wires. In general, copper pair cables of various sizes are commonly observed in the distribution plant, especially in the line segments toward the local subscribers, although the telephone companies have been focusing more on the use of fiber cables because of their increased capabilities for providing advanced services (high-speed digital transmission). Figure 4.10 shows typical copper pair cables, including a self-supporting type with an integrated steel messenger.

Figure 4.11 Jacketed and unjacketed coaxial cables. Source: Courtesy of CommScope®.

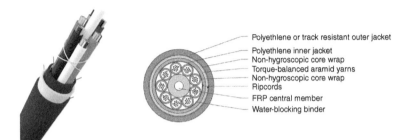

Polyethlene or track resistant outer jacket
Polyethlene inner jacket
Non-hygroscopic core wrap
Torque-balanced aramid yarns
Non-hygroscopic core wrap
Ripcords
FRP central member
Water-blocking binder

Figure 4.12 All dielectric self-supporting fiber cable. Source: Courtesy of AFL.

Although coaxial cables were originally used by the cable industry to provide cable television (CATV) service, they now also provide Internet and other communications services. Thus, whereas the "T" in the SCTE had previously indicated "Television," it more recently had been updated to indicate "Telecommunications" – i.e. the Society of Cable *Telecommunications* Engineers. Figure 4.11 shows a typical coaxial cable which may, or may not, have a protective jacket.

Fiber cables are available in sizes ranging from only a single fiber to hundreds of fibers. Fiber cables may be directly lashed to a messenger, similar to the other type cables, but self-supporting types are readily available, including in a figure-8 configuration, or with a central strength member, as illustrated in Figure 4.12. All-dielectric self-supporting (ADSS) fiber cables are also available, which include

(a)   (b)   (c)

Figure 4.13 Wireless antennas mounted on utility poles. (a) Pole-top mounting. (b) Sidearm/crossarm mounting. Source: Courtesy of ConcealFab Corporation. (c) Strand mounting. Source: Courtesy of Wade4Wireless.

nonmetallic strength members, to eliminate vulnerability to the effects of power contacts or lightning.

4.5 Wireless Attachments

Wireless (cellular) facilities are increasingly important communications attachments on utility poles. In particular, the proliferation of 5G networks and small-cell technologies provide a challenge to find sufficient mounting locations for the antennas, in order to provide adequate coverage, for which the ubiquitous utility poles are an obvious target. Figure 4.13 shows antennas (and support equipment) mounted at three different locations on the pole, including at the top, on a sidearm/crossarm, or on the strand. **Appendix B** provides a brief discussion of related design and safety issues in the placement of antennas on the poles.

5

Cable Installation

5.1 Conductor and Cable Placement

Differences between the electric supply and communications lines, as described in Chapter 4, can affect the method of installation of the conductors and cables. Whereas power conductors are typically directly suspended between insulator pins on horizontal crossarms or vertically mounted insulator brackets, located on adjacent poles, most communication cables are continuously lashed to a previously independently installed steel messenger wire. Communication cable designs that have an integrated messenger, or strength member, eliminate the step of installing a separate messenger, and the subsequent lashing operation. Supply cables may also use messengers to provide support by a spacer bracket, as commonly implemented in the Hendrix-type system (Figure 4.6), although supply cables are sometimes also lashed directly to a messenger.

There are two basic methods for installing aerial cables: stationary reel and moving reel. As suggested by the terminology, the cable reel may remain at one end of the pole line segment, or it may be mounted on the truck as it traverses the route, as the cable is temporarily placed on the ground or along the pole supports. Figure 5.1 illustrates the stationary reel method, in which a communications cable is fed off the reel located at one end of the segment, and the cable pulled and temporarily suspended beneath the existing or previously installed messenger wire. A somewhat similar method may be used for supply conductors, which may be slid along the ground, or suspended in the air, not touching the ground (Shoemaker 2017). The conductors may then be placed on insulators on the poles.

The stationary reel method is also used for placing the Hendrix-type electric supply cable, but in this case, multiple conductor reels are used, such as illustrated in Figure 5.2 (Western 2013).

Figure 5.3 shows the moving reel method in which a communications cable is fed off the reel on the moving truck. In both cases (stationary reel and moving reel), the communications cable will be lashed to the messenger, but not necessarily

Overhead Distribution Lines: Design and Application, First Edition. Lawrence M. Slavin.
© 2021 The Institute of Electrical and Electronics Engineers, Inc.
Published 2021 by John Wiley & Sons, Inc.

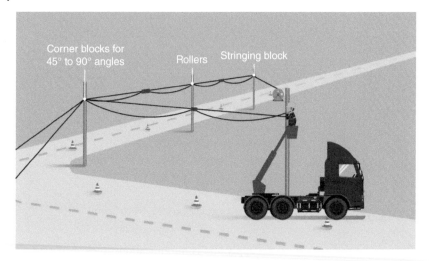

Figure 5.1 Stringing overhead communications cable – stationary reel method. Source: Courtesy of OFS Fitel, LLC.

proceeding in the same direction (see Section 5.2). The moving reel method may also be used for supply cables for which the conductors are laid on the ground, avoiding a dragging motion that may cause damage, prior to placement on the insulators.

5.2 Lashing Operation

Figure 5.4 shows a cable lasher, with the direction of travel as indicated. The front of the unit scoops up the cable and allows it to be lashed to the existing messenger. The same procedure can perform an "overlash" in which the new cable is lashed to an existing cable/messenger combination that had been previously lashed together. Overlashing procedures are commonly employed by communications companies as a convenient, efficient means of adding new technologies or upgrading their networks, as described in Section 5.3.

For the stationary method, the lashing operation is typically performed in the opposite direction of the original pulling position, proceeding toward the reel position, as illustrated in Figure 5.5. However, for the moving reel installation, the lashing operation is usually performed simultaneously with the truck and reel movement, proceeding in the same direction along the span, as shown in Figure 5.6.

As described, both methods assume an existing or previously installed messenger strand. In practice, the messenger may also be installed similar to either the stationary or moving reel method, or by any other convenient procedure.

Source: Courtesy of Marmon Utility LLC

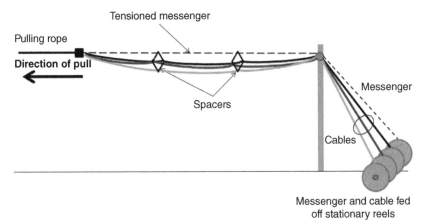

Figure 5.2   Stringing Hendrix-type supply cable – stationary reel method.

5.3   Overlashing

"Where applicable, consider using *overlash* construction methods rather than placing new strand."

[emphasis added]

- TELCORDIA *BLUE BOOK - MANUAL OF CONSTRUCTION PROCE-DURES*[1]

This statement is based upon the practical and physical advantages associated with the reuse of an existing strand for the installation of new communications cables, as often implemented by a communications utility to meet its additional or growing needs. Figure 5.7 illustrates the geometric efficiencies of forming a single compact bundle rather than attempting to place new cables on separate

---

1 Make-Ready Survey Checklist (Telcordia 2017).

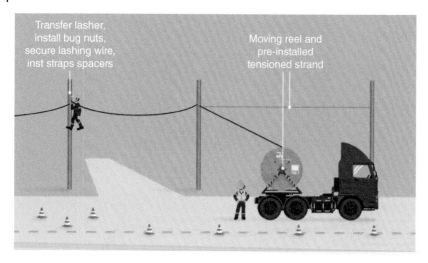

Figure 5.3 Stringing overhead communications cable – moving reel method. Source: Courtesy of OFS Fitel, LLC.

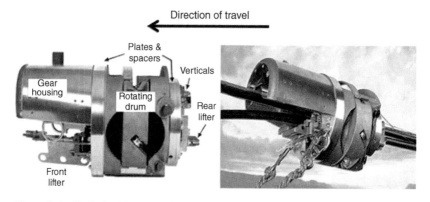

Figure 5.4 Typical cable lasher. Source: Courtesy of General Machine Products (KT), LLC.

strands.[2] The latter would require significantly more pole space, and impose significantly greater loads on the supporting structure, than that of the overlashed bundle. Additional details regarding the spatial and physical characteristics, including the effects on pole and strand loading, as well as sags and clearances, are provided in Chapters 6 and 7.

---

2 Based on SpanMaster® Release 3.1, Commscope Inc.

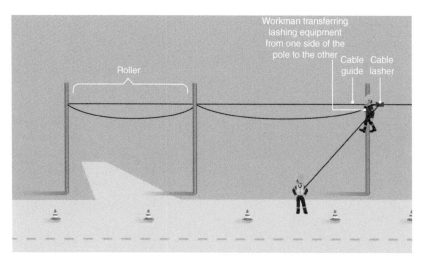

Figure 5.5 Lashing communications cable – stationary reel method. Source: Courtesy of OFS Fitel, LLC.

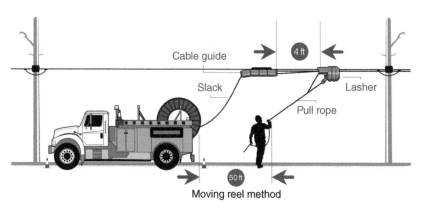

Figure 5.6 Stringing and lashing overhead communication cable – moving reel method. Source: Courtesy of LANshack.com.

While overlashing offers obvious benefits, the practice does impede the removal of older cables. In some cases, the older cables may be reactivated in the future, if/when necessary, and should therefore remain in place. However, some older cables may be permanently deactivated – and therefore considered "abandoned."

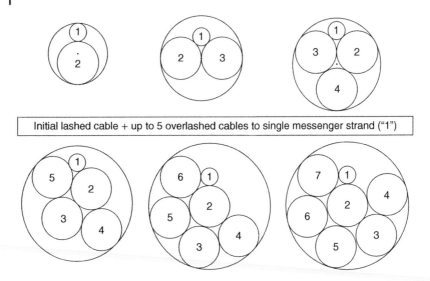

Initial lashed cable + up to 5 overlashed cables to single messenger strand ("1")

Figure 5.7 Overlashed cable bundles.

The continuing presence of such cables, however, is usually of little of no consequence, except to the degree indicated in Chapters 6 and 7. It is suggested that the removal of such cables should not be aggressively pursued because of the significant potential for damaging existing cables, and causing unnecessary service disruptions, while attempting to resolve a dubious problem. The NESC, for example, recognizes that such cables are generally not a hazard.[3]

---

3 NESC Rule 214B3: Lines and equipment permanently out of service shall be removed *or maintained in a safe condition* [emphasis added] (NESC 2017).

6

NESC® Requirements (Strength and Loading)

6.1 National Electrical Safety Code (NESC)

010. Purpose

A. The purpose of the NESC is the practical safeguarding of persons and utility facilities during the installation, operation, and maintenance of electric supply and communication facilities, under specified conditions. Accredited Standards Committee C2-2017 *NATIONAL ELECTRICAL SAFETY CODE*[1]

The National Electrical Safety Code (NESC) is applicable to outdoor lines under the control of the utilities, containing basic safety requirements for both overhead and belowground power supply and communications facilities. The Code is reissued every five years and either the latest or a specified previous edition is "adopted" by most public service or utility commissions, or reflected in industry requirements, in the United States, including its territories (see Chapter 9). Although not intended as a "design specification or as an instruction manual," many of the requirements directly or indirectly impact the design and construction, as well as utilization, of the facilities. In addition to the rules for electric supply stations, underground lines and work rules provided in other parts of the NESC, Part 2 contains safety rules for the overhead lines. The latter comprise detailed strength requirements and clearance rules for the support structures (e.g. poles) and cables or messenger wires. The physical characteristics

---

1 The rules described in this manual are based on the 2017 edition (NESC 2017), but are presented in a manner consistent with possible changes in the next edition. However, Rule 013B (Existing Installations) outlines the rules that are appropriate for existing installations, based on the time of initial construction and history of the line, which rules may therefore differ somewhat from those described herein.

Overhead Distribution Lines: Design and Application, First Edition. Lawrence M. Slavin.
© 2021 The Institute of Electrical and Electronics Engineers, Inc.
Published 2021 by John Wiley & Sons, Inc.

of the suspended wires, conductors, and cables directly impact the design and construction practices in order to comply with these requirements.

6.2 Loading Requirements

Section 2.3 discusses the effects of loads on the pole structures. In particular, the strength of the pole, as well as crossarms and various support hardware, must be compatible with the basic loading conditions specified by the NESC, depending upon the grade of construction, as described in Section 6.3. For typical joint-use applications, the required loading condition is that of NESC Rule 250B, corresponding to the storm loads of the geographic districts illustrated in Figure 6.1. (The term "Zone" had been inadvertently inserted into Figure 6.1 and will be corrected in a future edition.) These include the three districts designated as Heavy, Medium, and Light storm loads, as well as that associated with Hawaii and the other "Warm Islands." (A similar map, with corresponding zones, is specified for the purposes of determining sags and clearances; see Chapter 7.)

The Rule 250B storm loadings vary from $1/2$-in. radial ice thickness in the Heavy loading district to $1/4$-in. radial ice in the Medium loading district, to the absence of ice in the Light loading district. These ice conditions, furthermore, are to be

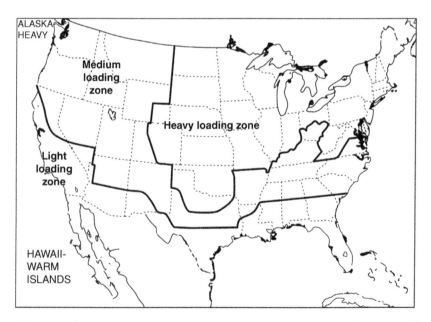

Figure 6.1 Storm loading districts for determining strength (Rule 250B). Source: NESC, C2-2017.

Table 6.1 Rule 250B storm loadings.

	District		
	Heavy	Medium[a]	Light[b]
Ice thickness (in.)	½	¼	0
Wind pressure (psf)	4	4	9
Temperature (°F)	0	15	30
Additive constant (lbs/ft)	0.30	0.20	0.05

a) Applies to Warm Islands, above 9000 ft altitude.
b) Applies to Warm Islands, below 9000 ft altitude, with temperature = 50 °F.

considered in the presence of wind pressures of 4, 4, and 9 psf, respectively, indicating that the heavy/medium/light designations only refer to the ice thickness, and not the wind pressures. Since, in practice, the dominant load on the support structure (pole) derives from the lateral wind pressure, the "light" loading area may experience the greatest calculated loads on the structure, depending on the wire or bundle[2] diameter. Although not stated, the wind pressures are sometimes interpreted as corresponding to wind speeds of 40 and 60 mph, respectively (see Section 6.6).

Table 6.1 provides the required Rule 250B storm loadings, including the temperature, which, in addition to the ice weight and wind pressure, affects wire tensions. Table 6.1 also includes the "additive constant," used only for the purpose of determining wire tension in the NESC, as explained in Section 6.4 and illustrated in Figure 6.5.

The wind pressure is applied to the overall diameter of the ice covered wires, cables, and conductors, including messenger, or overlashed bundle, as appropriate, as shown in Figure 6.2. (The ice covering need not be applied to the structure or wire supports nor to supported equipment or hardware.) The corresponding forces, acting transverse to the line, then typically become the dominant factor in determining the required strength of the structure, for tangent lines. Figure 6.3 illustrates the associated bending moment at the groundline imposed by the lever arms corresponding to the heights of the attachments on the pole, acting as a cantilever (see also Section 2.3).

The overall diameter of the ice-covered group of wires or conductors is given by

Overall (ice-covered) diameter = Bundle diameter + 2 × Radial ice thickness

---

2 In this manual, the term "bundle" refers to a group of cables and/or wires that are lashed together, as described in Sections 5.2 or 5.3. This differs from the terminology of the NESC in which a bundled conductor refers to the Hendrix-type supply cable, or equivalent.

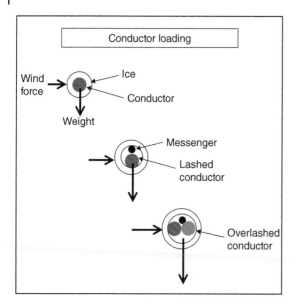

Figure 6.2 Storm loading on ice-covered bundle.

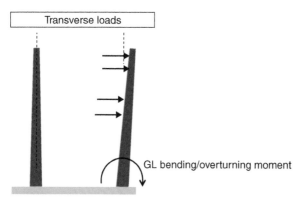

Figure 6.3 Transverse loads applied to pole acting as a cantilever, resulting in groundline bending/overturning moment.

and the weight of the ice-covered group of wires or conductors, including that of the wires themselves, may be approximately determined as

Ice-covered weight (lbs/ft) = Total weights of wires (lbs/ft) + 1.24

× Radial ice thickness (in.) × [Bundle diameter (in.) + Radial ice thickness (in.)]

Although the effect of the weight of the structure or supported facilities, including ice, is usually not a significant load in determining the strength of the pole compared to the effect of the transverse (horizontal) loads acting to bend the pole as a cantilever, the wire tensions are directly affected by the weights of the cables, including ice covering. For tangent lines, these wire tensions act in opposite (longitudinal) directions at the pole and usually have no significant impact on the required strength. (For unequal spans, or unequal ice loads, possible differential tensions may be transmitted to the structure, but the flexural/stiffness characteristics of the pole tends to allow some degree of load equalization.) However, at line angles, the wire tensions vectorially add to the loads on the structure, as illustrated in Figures 2.5 and 2.8, and are usually the dominant load in such cases. Wood poles at significant line angles are typically supported by guys, as described in Section 2.5, which strength is also specified in the NESC. Engineered (e.g. steel, concrete, laminated wood, etc.) poles may be designed with sufficient strength to withstand the effect of tensions, plus wind loadings, at significant line angles, without the aid of guys. In general, the strength of the pole structures and systems must be compatible with the imposed loads, and the load factors and strength factors, as discussed below. The NESC also provides limits on the allowable tensions in the wires or conductors themselves.

6.3 Strength Requirements

The requirement that the structure or other component be able to withstand the imposed loads may be expressed by the relation:

Strength factor × Strength ≥ Effect of (Load factor × Load)

The introduction of the load factor (LF) and strength factor (SF) is consistent with the load and resistance factor design (LRFD) format, which accounts for material variability and desired design margin. The strength (resistance) factor is typically ≤1.0, depending upon the basis on which the strength is specified. For example, the ANSI O5.1 standard specifies an average fiber strength for the most commonly used wood species (Southern Pine, Douglas Fir, Western Red Cedar). In general, it is not appropriate to base a design only on the average strength of a wood pole because of the possible significant material variability of a naturally grown wood pole; see Section 2.2. Therefore, a strength factor that "derates" the material to provide a more conservative value is used to help enhance safety. However, for engineered products, such as steel or fiber-reinforced polymer, with relatively low variability and for which a "minimum" strength may be provided, a strength factor of 1.0 may be used – i.e., no derating is required. Thus, the NESC allows a strength factor of 1.0 to be applied to the fifth percentile strength (i.e. 5% lower exclusion limit) for fiber-reinforced polymer structures and crossarms.

In addition to the material, the specified strength factor depends upon the grade of construction. Grade B is the highest level defined in the NESC and is applicable for lines crossing railroads, limited-access highways, and navigable waterways requiring crossing permits, and some high voltage situations (exceeding 22 kV).

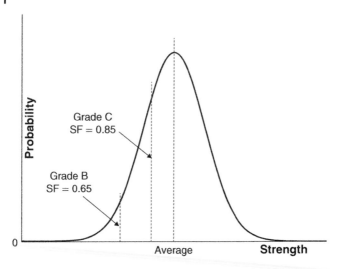

Figure 6.4 Strength factors applied to average strength of wood pole (Rule 250B).

Grade C is applicable for most applications with primary power (i.e. exceeding 750 V) on the pole. Grade N applies to structures supporting only secondary power (0–750 V) or communications lines. Grade N also applies to emergency or temporary structures.

Figure 6.4 illustrates the implications of the Grade B and Grade C levels as applied to wood structures, as specified in the NESC for the Rule 250B loads. For an assumed coefficient of variation of 20% (ANSI 2017), the strength factor of 0.65, for Grade B construction, results in an assumed wood pole strength less than the 5% lower exclusion limit – i.e., more than 95% of the poles in the selected class (Section 2.2) will have a strength equal to, or exceeding, the derated strength. For example, the assumed strength of a Class 3 wood pole would be $0.65 \times 3000\,\text{lbs} = 1\,950\,\text{lbs}$. In comparison, the strength factor of 0.85, for Grade C construction, results in a derated wood pole strength of 2550 lbs, corresponding to a lower exclusion limit less than 25%, exceeded by more than 75% of the poles in the selected class.

There are no analogous quantitative requirements for Grade N construction, which "need not be equal to or greater than Grade C," but poles must be able to "withstand expected loads," including working linemen. One logical design approach for Grade N construction would be to require the strength to be compatible with the storm loads of Rule 250B, but with both the load and strength factors equal to unity, and possibly somewhat stronger for temporary structures otherwise required to be of higher grade construction. In recognition that the NESC does allow Grade N construction for many commonly encountered applications – including communications-only poles, and/or those with secondary power (e.g. pole lines in the absence of transformers) – the Telcordia *Blue Book* notes that it is normally reserved for exclusive private-rights-of way, service drops, and emergency or temporary installations (Telcordia 2017).

The load factor in the above formula is typically ≥ 1.0 and is intended to amplify or modify the magnitude of the specified loads, which, similar to the strength factors, is intended to help enhance safety. The load factors are specified as a function of the type or direction of the load – i.e., vertical (weight), transverse (wind vs. wire tension), or longitudinal (general vs. dead ends) – and are again based on the grade of construction.

Ideally, the load factors would only depend upon the type load (and grade of construction), while the strength factors would be only dependent upon the material (and possibly grade of construction). Nonetheless, the load factors in the NESC do somewhat reflect the type material for some structural items, and the strength factors are somewhat dependent upon the type storm loading for some materials. In addition, there are two categories of load factors for Grade C construction for the Rule 250B loads, for the critical transverse wind load – i.e., a factor (1.75) specified for typical structures, and a larger factor (2.2) for the less common case of structures supporting wires crossing over lines below. Table 6.2 shows the load

Table 6.2 NESC load and strength factors for Rule 250B district loadings.

		Grade B	Grade C
Load factors	Vertical	1.50	1.90 – wood[a]
	Transverse		1.50 – metal[b]
	Wind	2.50	**1.75** (2.20 – wire crossings)
	Wire tension	1.65	1.30[c]
	Longitudinal		
	In general	1.10	No requirement
	At dead-ends	1.65	1.30[c]
Strength factors	Wood[a]	0.65	**0.85**
	Metal[b]	1.0	1.0
	Support hardware	0.9	0.9
	Guy wire	0.9	0.9
	Guy anchor and foundation	1.0	1.0

a) Also includes reinforced concrete.
b) Also includes pre-stressed concrete and fiber-reinforced polymer.
c) A load factor of 1.10 applies to metal or pre-stressed concrete portions of structures and crossarms, including guys and anchors.

and strength factors specified by the NESC for use with the Rule 250B storm load-ings on distribution poles. These factors apply to the structures themselves and associated structural components, such as crossarms, guys, foundations, etc.

Although a safety code, the NESC does not use the term "safety factor" relative to the strength and loading criteria. However, based on the formula above, it may be seen that the ratio of the load factor to the strength factor corresponds to an effec-tive safety factor.[3] For wood poles, the ratios are approximately 2-to-1 (=1.75/0.85), the familiar "safety factor" for typical distribution (Grade C) applications, and approximately 4-to-1 (=2.50/0.65) for the more limited Grade B applications.

6.4 Wire Tensions

The NESC also provides strength requirements for wires and conductors, includ-ing messengers, but which do not differentiate between Grade B and Grade C. The peak tension under storm load conditions is determined using the ice loads, wind pressures, and temperatures, as well as the "additive constants" of Table 6.1 for the Rule 250B loads, as illustrated in Figure 6.5. The additive constants arose as a historical artifact introduced in the NESC when the specified wind pressures were significantly reduced, but for which it was desired to maintain the previous ten-sions and sags (clearances), which were determined as a result of the Rule 250B district loads. The reduced wind pressures for structural loads were offset by other changes, including reduced allowable wood stresses. The additive constant, how-ever, is not (directly) considered for the purpose of determining the vertical or

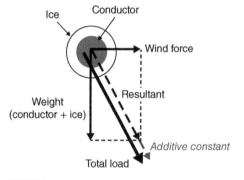

Figure 6.5 Conductor loads for determining wire tension.

---

3 For nonlinear (e.g. "*P*-delta") considerations, such as may be reflected in some design procedures, the simple ratio of the load factor to the strength factor may not represent the effective "safety factor," since the load factor may have a disproportionate influence on the required strength. This is because the term "Effect of (Load Factor × Load)," in the formula for the strength requirement, is not necessarily equal to the "Load Factor × (Effect of the Load)" for nonlinear phenomena.

transverse loads transferred to the structures or line supports. Since the use of the additive constant is sometimes considered to be an unnecessary complication, this component may possibly be eliminated in a future edition of the NESC.

Unlike the strength requirements for supports, which are based on the load and strength factors, the wire or conductor tensions, including messengers, are stated as not exceeding 60% of their rated breaking strength (RBS) at the Rule 250B storm loading. Chapter 8 discusses a method for determining the tension resulting from applied loads, such as that illustrated in Figure 6.5, for wires or conductors that behave elastically, including typical steel messengers. Commercial software tools are available that also include the effects of inelastic behavior (Section 8.2).

NESC Rule 261H provides an additional requirement for supply conductors to help avoid or mitigate the effects of Aeolian vibration. The initial tension is therefore limited to 35% RBS, immediately following installation and without external loading, and 25% RBS at "final tension," without external loading, which includes the effects of inelastic deformation.

For Grade N conductors, which nominally includes most communications and secondary power lines, there are minimal strength requirements. For such supply lines, some conductor sizes are specified, and the tension of the service drops conductors must not exceed the strength of the attachment or support under the unspecified "expected loads." For most communications lines, there are no formal requirements, but the 60% tension limit, under the Rule 250B storm loads, is usually observed, except possibly for exclusive private-rights-of way, service drops, and emergency or temporary installations.

6.5 Guyed Poles

Depending upon the pole stiffness or rigidity, guys used to meet the strength requirements may be considered to act as an integral part of the structure, or, conversely, to counteract the entire (transverse or longitudinal) load in the relevant direction. Relatively rigid structures, such as steel or concrete, will share the effect of the loads with the supporting guy, if any. In comparison, relatively flexible structures, such as wooden poles, or those constructed of fiber-reinforced polymer, will deflect sufficiently until the steel guy wire absorbs most or all of the intended load. For the latter structures, it is therefore assumed that the guy will support the entire load of interest. Figure 6.6 illustrates the guy supporting the storm load, including that from the wind pressure on the pole and (ice-covered) wires, as well as the potentially large unbalanced load at a line angle (see Section 2.5) resulting from the messenger or wire tension, which tension is also increased by the weight of ice and wind pressure acting on the ice-covered wires, as appropriate. Guys may also be advantageously employed in the absence of line

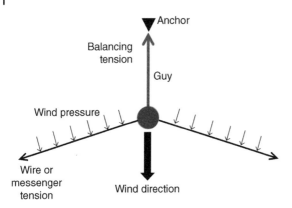

Figure 6.6 Guy supporting storm load.

angles, as may be necessary in some cases, to support the transverse load resulting from the cumulative wind forces acting on the pole and attachments, or if there are foundation (stability) issues.

Although anchor guys will counteract the intended load, they will also place an additional load on the pole, acting in the vertical/axial direction. This additional load will tend to be large, as discussed in Sections 2.5 and 2.6, and the possibility of buckling as a column (Figure 2.9) should be considered.

6.6 Extreme Wind Loads ("60 ft Limit")

In addition to the Combined ice and wind district loading (Rule 250B), the NESC requires consideration of two other storm conditions: Extreme wind (Rule 250C) and Extreme ice with concurrent wind (Rule 250D). These "extreme" loads correspond to relatively rare events (e.g. mean return interval of 50 years or greater), characterized by very high winds (e.g. hurricanes) or very heavy ice loads. These additional requirements, however, are only applicable for limited applications, most notably for cases in which some portion of the structure, or supported facilities, exceeds 60 ft above ground (or water) and has been understood to be a proxy to distinguish between transmission structures and the typically shorter distribution poles. While not mandatory, these load conditions have sometimes also been applied to other cases, where the additional strength criteria are believed to be of potential benefit, and the requirements may possibly be extended to a limited set of categories in the future. (**Appendix C** provides more details on Rules 250C and 250D.)

The familiar 60 ft height criteria is primarily based on wide-spread industry experiences which indicate that the practical strength requirements under these conditions are not effective in limiting damage to the latter structures. These weather conditions are accompanied by falling branches and/or flying debris which impact the lines, resulting in high wire tensions, often breaking the distribution poles. Transmission structures are less vulnerable to these detrimental effects since their lines often are above the height where falling vegetation and wind-blown debris are more commonly encountered. Nonetheless the strength required to withstand the district loads of Rule 250B, consistent with the load and strength factors, and other conservative assumptions described below, as well as the maintenance requirements (Section 6.7), result in a relatively robust system that tends to perform well, even in the presence of the extreme wind and ice loads, as may be directly imposed on the distribution lines. In fact, depending upon the geographic area and conductor sizes, the strength required to withstand the Rule 250B loads, with corresponding strength factors, may exceed that of the Rule 250C and Rule 250D loads, especially for Grade B construction.

Unfortunately, this background is not always understood, and the 60 ft limit may incorrectly be characterized as an unreasonable "exemption," with no technical justification. In fact, such distribution poles have been compared to free standing light poles, which may be required to withstand extreme wind loads, similar to those of Rule 250C. This is, however, an inappropriate analogy, since light poles are not appreciably exposed to the falling branches or flying debris that impact the distribution lines spanning utility poles, which wires act as a net capturing the vegetation and other free moving objects. Indeed, Rule 261A recognizes this distinction and does require the poles to withstand the extreme wind loads of Rule 250C, prior to the installation of the wires spanning the structures:

> *All structures including those below 18 m (60 ft) shall be designed to withstand, **without conductors**, the extreme wind load in Rule 250C applied in any direction on the structure and any supported facilities and equipment which may be in place **prior to installation of conductors**.* [emphasis added]

In addition to the relatively conservative load and strength factors, another reason for the good performance of the distribution poles, including during many extreme events, is that the required wind pressures are generally assumed to be applied at right angles to the lines, whereas the actual wind direction may not be that severe. For example, the following simplified formula applies to a wind impacting a wire line at right angles, confirming that the 4 and 9 psf wind pressures are consistent with 40 and 60 mph wind events, respectively, for an assumed shape factor of 1.0.

$$\text{Wind pressure } \left(\text{lbs/ft}^2\right) = 0.00256 \times \left[\text{Wind speed (mph)}\right]^2 \times \text{Shape factor}$$

In most cases, however, these wind speeds will impact the lines at a lower (less than 90°) angle, Θ, significantly reducing the applied load – by a factor of $\sin^2 \Theta$

(ASCE 2020). Conversely, and more significantly, the required 4 and 9 psf wind pressures correspond to significantly higher wind speeds that may impact at the less severe angles. Furthermore, the tensioned wire lines connecting the poles may allow a weaker than average pole – which is the reason for requiring the strength (reduction) factors – to benefit from support provided by potentially stronger poles located at the end of the adjacent spans, as the weaker pole typically laterally deflects to a greater extent than the adjacent (stronger) poles. Older editions of the NESC allowed formal consideration of this effect, referred to as the "Average Strength of Three Poles," under Rule 261.

In those instances where it is necessary to consider the extreme loadings of Rules 250C and/or 250D, the calculations of the required storm loads are more complicated than that of the district loadings, but have generally been incorporated into commercially available software tools (Section 6.9), as an option.

6.7 Allowable Deterioration

Assuming the poles meet the basic strength criteria of Rule 250B (Combined ice and wind loading), required for all structures, it is generally acknowledged that the most cost-effective strategy for maintaining a robust distribution network is a combination of vegetation management, as suggested above, and effective maintenance practices (see Section 3.3). In order to help maintain most or all of the required basic strengths of the structure, NESC Table 261-1 limits the allowable deterioration. Thus, for engineered materials (i.e. metal, fiber-reinforced polymer, or prestressed-concrete), there is no allowed reduction in the required strength, while a reduction of 1/3 (33%) the original required strength is allowed for wood (and reinforced-concrete) materials, for the Rule 250B loads. Poles not meeting this minimal strength level must be rehabilitated or reinforced. (Examples of reinforced poles are illustrated in Figure 3.9.)

The allowed reduction in strength must not be misunderstood to be a reduction in the required strength of the pole independent of any deterioration. For example, if an undeteriorated pole is at or near its capacity, there cannot be any additional attachments for which the total load would then exceed the basic strength, considering the original load and strength factors. However, if the same pole, with the same original attachments, deteriorates, consistent with the 1/3 allowance, the pole need not be replaced or rehabilitated at that time. Nonetheless, any pole showing decay should be monitored (inspected or tested) on a regular basis.

Whereas a finite amount of deterioration is therefore allowed, actually determining when a pole is at this point may sometimes be a challenge. The bending strength of a wood pole, for example, is traditionally determined using the simple formula in Section 2.2 to determine the allowable bending moment (prior to application of the strength factor), and is typically based on the outer circumference at the groundline and the designated fiber strength in the ANSI O5.1 standard. This procedure, however, assumes there is no significant internal decay. Although there are procedures and tools for attempting to determine the amount of internal decay in in-service poles, the calculation of the corresponding bending strength is very sensitive to the size, shape, and location of any estimated voids or decay. More sophisticated products are also available, but, at the time of this publication, the accuracy of such techniques for quantifying the remaining strength of the pole is not obvious; see also Section 3.3.

Even in the absence of internal decay, the use of the pole circumference of an aged pole, which may have suffered some exterior degradation over the years of service, may not be strictly technically correct for determining the degree of reduction in strength from that at initial installation of a new pole. This is because of designated fiber strength values, and corresponding class loads, actually represent effective or average values across a pole class, for which the poles fall within specified limits of circumference. The designated fiber strength is most appropriate if applied to the minimum specified size (circumference) within a pole class, and a somewhat lower value may be more appropriate if applied to an actual pole size. Thus, the calculated bending strength may then be somewhat overestimated, and a pole determined to satisfy the 1/3 requirement may actually be somewhat weaker. A possible alternate approach may therefore be to assume a fiber strength approximately 10% lower than the designated strength, to account for the pole oversize within each class. Using this procedure, initially smaller poles within a class size would result in a lesser allowable strength reduction than that of initially larger sizes within the same class, with an initially average size pole within the class corresponding to the 33% allowable reduction when determining its required strength.

6.8 Overlashed Cables

The overlashing option described in Section 5.3 provides distinct benefits compared to installing separate, individually supported cables, including reduced loads on the pole, as well as more efficient utilization of vertical pole space, as discussed in Chapter 7. The geometry of a bundle of several cables, supported on the same messenger, is considerably more efficient with respect to transverse wind loading than that corresponding to the direct sum of the diameters of individual supported cables (i.e. vertically stacked). The default NESC model for determining the wind sail area is as follows (NESC Rule 251A3):

> *An appropriate mathematical model shall be used to determine the wind and weight loads on ice-coated conductors and cables. In the absence of a model developed in accordance with Rule 251A4, the following mathematical model shall be used:*
>
> a. *On a conductor, lashed cable, or multiple-conductor cable, the coating of ice shall be considered to be a hollow cylinder touching the outer strands of the conductor or the outer circumference of the lashed cable or multiple-conductor cable.*

This model is also considered to apply to a bundle of several (overlashed) cables, for various radial ice thicknesses, including in the absence of ice.

As a result of the efficient bundle geometry, the incremental sail area (and wind load), including ice thickness, associated with each additional (overlashed) cable is small compared to that of each cable itself – and especially compared to the alternative of each cable supported on a separate strand. Figure 6.7 compares the typically critical transverse wind loads for multiple cables as installed on separate messenger strands, or in a bundle on the same strand,[4] for the Heavy loading of Rule 250B. The illustrated results are based on a commonly used 6M (5/16-in.) messenger strand, and nominal 0.75-in. diameter cables, representative of typical coaxial or fiber cables, consistent with the bundle configuration illustrated in Figure 5.7, with corresponding overall bundle diameters. The wind load is also shown if the sail area is calculated by simply assuming the cables are vertically stacked on the same strand and adding the diameters. Although the quantitative results will vary depending on the actual cable sizes, and the storm district (e.g. Heavy, Medium, or Light loading) for the area in question, the results are qualitatively similar.

The incremental wind force for each addition to an overlashed bundle is seen to be only a small fraction of the prior load, and in some cases may be negligible, due to the nesting of the additional cable within the bundle of existing cables. This phenomenon, in addition to the relatively low attachment height on the pole for communication lines, is the reason that it is often unnecessary to perform detailed load analyses of a typical pole for an overlash of a fiber or coaxial cable. Indeed, it is quite possible that a particular pole loading analysis, following an overlashing, may result in the pole seemingly having *greater* margin available than prior to the addition, because of possible differences in analytical methods and assumptions employed by different analysts.

The overlashing option also minimizes the loads imposed on poles due to wire tensions at dead-ends or line angles, in comparison to the use of separate strands, for which each strand may have a significant initial tension. The determination of

---

4 The results in Figures 6.7 and 6.8 are based on SpanMaster® Release 3.1, Commscope Inc.

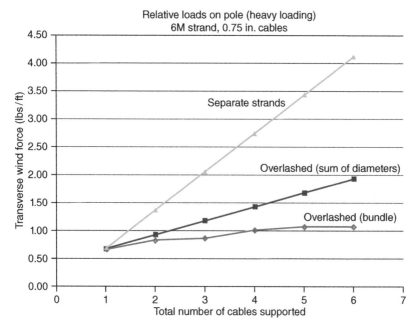

Figure 6.7 Comparison of transverse wind loads for multiple cables (Rule 250B).

the messenger tension is considerably more complicated, and less intuitive, than that of the transverse wind load, and depends upon the initial installed tension, temperature, and physical properties of the wire, including elastic modulus (stiffness) and thermal expansion characteristics. Figure 6.8 illustrates typical results, including the relatively small incremental storm tensions induced in the single messenger strand for each cable addition, which can be seen to be much lower than the sum of tensions corresponding to separate messengers, each supporting only a single cable. The results are similar for the various storm districts, messenger strands (6M, 10M, etc.), and other possible initial stringing tensions. Section 7.5 discusses the effect of overlashed cables on peak (midspan) sag and clearances.

In general, the wire tensions at dead-ends or line angles will be balanced by a system of guy wires and anchors, which system will therefore be greatly simplified by the overlashing alternative. Figure 6.8 also indicates that the magnitude of the increased tension in the strand, for multiple cables supported on the single strand, will generally be well below the corresponding 60% limit specified in the NESC for the Rule 250B loads.

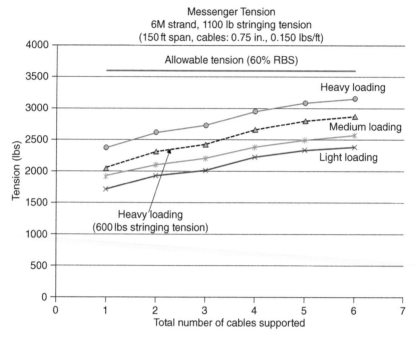

Figure 6.8 Incremental storm tension for multiple overlashed cables (Rule 250B).

6.9 Software Tools and Pole Loading Analysis

For many practical applications – such as tangent structures, or simple guyed configurations, without lateral cables or service drops – it is feasible to account for the various loads and compare to the strength of the pole system using a spreadsheet. However, the analysis becomes more difficult, or requires more experience, for more general cases that may include multiple guys, cables or wires attached at various angles, line angles with unequal spans, or other complications, including miscellaneous mounted equipment. Poles with multiple guys may be "statically indeterminate," such that pole deflections should be considered. At line angles, or other cases where the wire tensions are not balanced, it is necessary to determine wire tensions under storm loading. (Chapters 7 and 8 discuss wire sag and tension.) For cases where the extreme wind and ice loads of Rules 250C and 250D are applicable, the software tools also simplify the determination of the wind pressures on the wires, structures, and equipment.

In general, it is therefore convenient, and often necessary, to take advantage of the commercially available software tools that have been developed in order to perform an appropriate pole analysis. These products vary somewhat, but typically are based on the finite element method (FEM) for modeling the structure, incorporating the pole strength and stiffness characteristics, which also enable pole deflections to be considered. Such deflections allow nonlinear effects (e.g. *P*-delta), to be taken into account. The ability to determine the enhanced effect of pole deflections on the effective pole bending moments may also provide an indication of the potential instability ("buckling") of the pole; see Section 2.6. In some cases, the wire attachments are also modeled, allowing the tensions to be determined as part of the analysis, for which the pole deflections may also be important.

From a practical perspective, however, it should not be expected that the ubiquitous distribution poles, with their myriad of attachments, and variable and changeable nature, can be as precisely analyzed as the more stable, well-controlled, extremely critical transmission facilities, including lattice towers, *H*-frames, or single poles. Whereas the well-defined transmission lines may be accurately modeled, distribution poles include a variety of equipment, cables, and wires, including service drops, which may not always be conveniently accounted for, especially sags and tensions which depend upon the initial installation (stringing) conditions, and are subsequently temperature and weather dependent (see Sections 6.2 and 6.4). In general, therefore, it is neither practical nor cost-effective to attempt to determine the precise status of every distribution pole. Nonetheless, some analyses should be performed, as appropriate, as part of an asset management program, and to obtain a baseline for typical poles, representing a basis for comparison, or to evaluate potential problems arising from relatively long spans and/or numerous attachments, especially for older poles (see Section 6.7).

While the available software programs are generally invaluable, and are required for performing a load analysis in many cases, there is the possibility of misuse or user error. Such errors are more likely when one is totally dependent upon the software, without any prior experience or knowledge of what is reasonable. Thus, Chapter 10 contains some relatively simple examples that illustrate the application of the loads and an evaluation of the approximate required strength of a pole, without the need for sophisticated software. The ability to perform such calculations manually (i.e. with a spreadsheet) is useful to provide a sanity check, as well as to confirm one is using the software correctly. In many cases, such calculations may obviate the need to perform more detailed studies at that time, especially upon the addition of a relatively small attachment, including a typical overlash, as discussed above.

7

NESC® Requirements (Clearances)

7.1 Clearances

Figure 7.1 shows a joint-use pole with primary (high-voltage lines) and secondary (low voltage, "local power lines") supply lines located in the upper portion of the pole in the "supply space," and communications cables and other facilities further down the pole, in the "communications space," closer to the midsection. Within the communications space, the incumbent local exchange carrier (ILEC) facilities are usually located at the bottommost position, consistent with the joint-use agreements, and third-party communications companies (e.g. "Cable" or "CATV") located above. The supply space is separated from the communications space by the communication worker safety zone (CWSZ).

The rules for separation between the various facilities, as well as clearances above roadways and from other surfaces, are provided in detail in the NESC. The relevant section on such clearance requirements is the most extensive and prescriptive of the various sections of the NESC, including associated conditions and numerous tables, which are too complex to be provided in full detail in this manual. However, the NESC rules for such separation may be considered as in two general categories, including those whose primary function is (i) to protect the public and (ii) to protect the utility workers.

7.2 Clearance Zones

The vertical clearances between suspended wires, and between suspended wires and surfaces below, are directly related to the amount of sag of the wires or cables. Since the sag of supply and communications cables is also dependent upon the weather conditions, possibly including past history, it is necessary to specify the appropriate weather conditions under which the clearances are determined. In general, the clearance rules are based on considering possible "worst-case"

Overhead Distribution Lines: Design and Application, First Edition. Lawrence M. Slavin.
© 2021 The Institute of Electrical and Electronics Engineers, Inc.
Published 2021 by John Wiley & Sons, Inc.

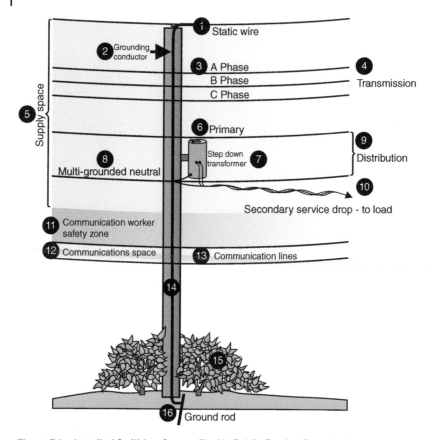

Figure 7.1   Installed facilities. Source: Florida Public Service Commission.

conditions, including "final sag" of wires and conductors (Section 8.2). For clearances above surfaces (including buildings), the wires are assumed to be ice loaded, at 32 °F, or at its maximum operating temperature, but a minimum of 120 °F. Their resulting sags are based on their physical characteristics, including stiffness and thermal expansion properties, and corresponding response to the specified loading and the initial (sag and temperature) installation conditions. These vertical clearances are determined in the absence of wind.

Figure 7.2 shows the four loading or weather Zones to be considered, which, for the present (and past) edition of the NESC, essentially match the Districts defining the loading requirements shown in Figure 6.1 and described in Section 6.2, for the purpose of determining the required strengths of structures and components. (Separate maps were introduced in the 2007 edition of the NESC to allow for the possibility of future differences in the storm loading conditions for the

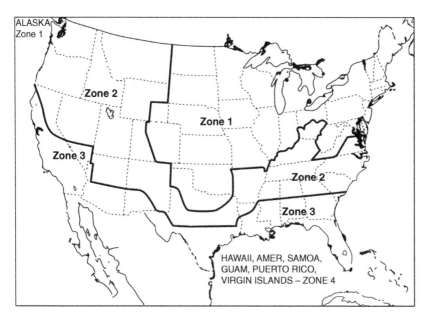

Figure 7.2   Clearance zones for determining clearances. Source: NESC, C2-2017.

Table 7.1   Prior storm loadings.

	Zone 1	Zone 2[a]	Zone 3[b]
Ice thickness (in.)	$\frac{1}{2}$	$\frac{1}{4}$	0
Wind pressure (psf)	4	4	9
Temperature (°F)	0	15	30
Additive constant (lbs/ft)	0.30	0.20	0.05

a)   Applies to Zone 4, above 9000 ft altitude.
b)   Applies to Zone 4, below 9000 ft altitude, with temperature = 50 °F.

purpose of specifying strength vs. those for the purpose of specifying clearance requirements.) The corresponding ice thicknesses are again indicated in Table 7.1. Table 7.1 also specifies other weather conditions (wind pressures, lower temperatures) and parameters (additive constant) that may initially appear to be unnecessary for the purposes of determining the related sags, since they do not agree with the immediate conditions under which the sags and clearances are determined. Indeed, except for the ice thickness, the prior storm conditions in Table 7.1 are not generally relevant for steel messengers, which are assumed to behave elastically, fully recovering from any prior storm loads. However, as discussed in Section 8.2, some wires, such as supply conductors with aluminum components, have a tendency to incur a permanent stretch, which may have occurred during the assumed previous ice and wind loadings, and/or from effects of long-term creep, and must be considered in any final sag calculations under the specified conditions (e.g. ice loaded, 32 °F, no wind, etc.).

7.3 Clearances Above Surfaces and Buildings

Table 7.2 provides typical minimum vertical clearances for distribution lines crossing over road surfaces and above buildings. (Caveats and exceptions to some of these values are provided in Tables 232-1 and 234-1 of the NESC.)

Table 7.2 Vertical clearances for typical distribution wires and cables (ft).

	Communication, messengers, supply neutrals	Secondary power (≤750 V), supply service drops (duplex, triplex, etc.)	Secondary power (≤750 V) open conductors	Primary power (>750 V to 22 kV) open conductors
Railroad tracks	23.5	24.0	24.5	26.5
Roadways, driveways, etc. subject to trucks (>8 ft)	15.5	16.0	16.5	18.5
Pedestrians, vehicles ≤8 ft only	9.5	12.0	12.5	14.5
Water areas[a]	14.0–37.5	14.5–38.0	15.0–38.5	17.0–40.5
Roofs not readily accessible to pedestrians	3.0	3.5	10.5	12.5
Roofs readily accessible to pedestrians	10.5	11.0	11.5	13.5
Roofs, ramps, etc. accessible to vehicles, but not trucks	10.5	11.0	11.5	13.5
Roofs, ramps, etc. accessible to trucks	15.5	16.0	16.5	18.5

a) Depends on water acreage (for sailboats).

Horizontal clearances from buildings vary from 4.5 to 7.5 ft, but significant reductions are allowed in some cases, including under wind displacement. Details are provided in Rule 234 of the NESC.

7.4 Clearances Between Wires

Whereas clearances above surfaces are intended to protect the public, clearances between the supply lines themselves, or between the supply lines and the communications lines (in the communications space) protect the utility workers. Of particular importance is the CWSZ, which allows sufficient space for the communication workers, who are generally not qualified to handle supply voltages, to safely perform their tasks. The CWSZ requires a minimum vertical separation of the familiar 40 in. at the pole (for supply voltages ≤8.7 kV), and 30 in. along the span, but which distances may be reduced to 30 and 22.5 in., respectively, for separation from a grounded neutral.

The worst-case conditions to be considered for maintaining the CWSZ correspond to the lowermost supply conductor being at the same worst case, final sag, conditions for specifying clearances above surfaces (i.e. ice and 32 °F, or elevated temperature), but the uppermost communication wire at the relatively low sag corresponding to the absence of ice and at the lowest anticipated consistent temperature (i.e. same ambient conditions). This combination results in the closest reasonable proximity of the wires in question.

Vertical clearances between conductors within the supply space must also be maintained at the conditions that result in the closest proximity of the wires in question, similar to that described for the CWSZ, but based on the lower wire without electrical loading. The minimum separation distances are a function of the voltage levels and ownership, as well as other details.

Although not inherently hazardous, vertical clearances are also specified between cables and facilities within the communications space. These include a 12-in. spacing where supported at the pole, and a 4-in. clearance along the span, which is generally understood to be maintained at a nominal 60 °F condition. These distances may be reduced if agreed by all parties involved (including the pole owner).

7.5 Overlashed Cables

The clearance requirements within the communications space, as well as those for the CWSZ, often may make it a challenge to physically fit additional separate messenger strands for each new cable required to satisfy the growing needs of subscribers, including new technologies. As described in Section 6.8, the use of

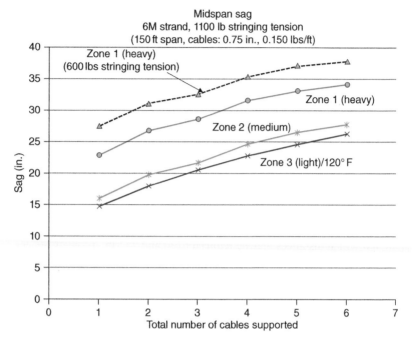

Figure 7.3 Comparison of midspan sags for multiple overlashed cables under ice or elevated temperature.

overlashing cables on an existing strand, presumably by the same owner, is a practical cost-effective alternative to adding additional messengers on a pole in order to support new cables. This is especially the case if the alternative requires replacement of an existing structure with a taller, usually higher strength pole, due to insufficient space above the present communications line without compromising the CWSZ. It is recognized that the greater sag characteristics of an overlashed bundle, if at the lowest position on the pole (typically the ILEC), may aggravate the ability to satisfy minimum clearances from surfaces and structures below. In particular, Figure 7.3 shows the midspan sags for overlashed cables, under the required worst-case conditions (i.e. ice loading or 120 °F).[1] However, the incremental sags are generally lower than the 12-in. spacing that may otherwise be required to place an additional (ILEC) messenger below the present lowest strand, and may well continue to meet required clearances. Alternatively, it may be possible to subsequently tauten the strand, with the overlashed bundle, to meet the clearance requirements, and still meet the maximum tension limits; see Figure 6.8.

---

1 The results in Figures 7.3 and 7.4 are based on SpanMaster® Release 3.1, Commscope Inc.

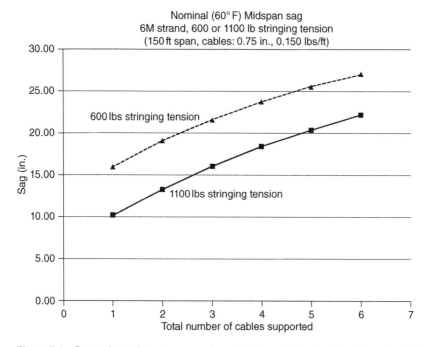

Figure 7.4 Comparison of midspan sags for multiple overlashed cables at nominal 60°F.

Clearances between vertically adjacent communication lines represent a different issue. The latter clearances are considered to apply at nominal installation conditions (60 °F), consistent with the sags indicated in Figure 7.4, and would apply, for example, between the third-party communications cables and those of the incumbent telephone company (ILEC), typically mounted below. Although the messenger spacing at the pole is specified as a minimum of 12-in., the clearance between cables/bundles along the span is allowed to be within 4-in. For situations in which multiple overlashed cables are added to an upper strand, this minimum separation may be compromised by the increased sag of the overlashed bundle. If necessary, the sag of the overlashed bundle may again be decreased by increasing the messenger tension, consistent with maintaining the peak storm tension within the allowable limit, or the required separation may possibly be accomplished by placement of spacers between the two sets of lines.

8

Principles of Wire Sag

8.1 Catenary

An understanding of the mechanical principles of suspended wires is important for the proper design and operation of overhead utility lines. For a cable or wire with little or no bending stiffness, and of uniform weight, the basic geometric shape for the suspended item is a catenary, characterized by relatively complicated mathematical hyperbolic (sinh or cosh) functions,[1] and the "catenary constant" (equal to the ratio of the horizontal component of tension to distributed weight). While it may sometimes be useful or necessary to utilize these mathematical representations for addressing specific cases, in most practical applications, for which the slope of the catenary is not excessive, the curve may be closely approximated by a parabola, for which the vertical location, y, along a suspended wire is related to the projected horizontal position, x, by the term $ax^2 + bx + c$, where a, b, and c are constants, with values depending upon the peak sag and coordinates of the end support points. In contrast to the hyperbolic functions, which often require iterative techniques to solve related mathematical equations, the parabolic representation and related equations are more amenable to mathematical solution.

Figure 8.1 illustrates the difference between the hyperbolic and parabolic representations, for supports at the same elevation and same catenary constant, as presented for an unreasonably large sag-to-span ratio of 20%, in order to be able to see a finite difference – less than 5% – in the results. Such a large relative sag is unrealistic, and may correspond to the wire resting on the ground below. A more realistic sag-to-span ratio of 2% or less would have a discrepancy of less than 1/20 of 1% – essentially negligible – and would not be discernible, either in the field or in a to-scale drawing. While larger discrepancies are possible where the support points are at significantly different elevations, such that the slope of the catenary is relatively high, the parabolic representation remains a convenient means for obtaining

---

1 The hyperbolic functions are defined as: $\sinh(x) = (e^x - e^{-x})/2$ and $\cosh(x) = (e^x + e^{-x})/2$.

Overhead Distribution Lines: Design and Application, First Edition. Lawrence M. Slavin.
© 2021 The Institute of Electrical and Electronics Engineers, Inc.
Published 2021 by John Wiley & Sons, Inc.

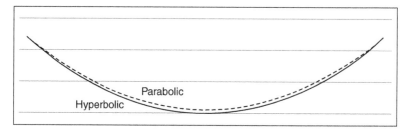

Figure 8.1 Comparison of hyperbolic and parabolic representations for suspended wire, for very large sag-to-span ratio.

practical estimates of the behavior of the wires under various conditions, including for the applications discussed below.

8.2 Initial and Final Sag

The terms "initial" and "final" are sometimes applied to sags or tensions, but the implied meaning is not always clear, and should therefore be interpreted by the context. Thus, the term "initial sag" in the NESC refers to the sag of a wire prior to the application of any external load, but may also be used, as in this chapter, to refer to the amount of sag prior to a change in this value to a subsequent value, for any reason. The term "final sag" is used in the NESC for specifying clearance requirements, referring to the sag of a wire, including any inelastic deformation caused by stretching under prior storm loads or long-term creep at more moderate conditions, when subject to subsequent specified ice loading and/or temperature conditions; see Chapter 7. The "initial tension" or "final tension" refers to that corresponding to the "initial sag" or "final sag" condition, respectively (see definitions in Glossary).

The effects of inelastic deformation are significant for many supply conductors, such as those containing aluminum components and should be considered for determining sags and tensions, as appropriate. These effects, however, are generally ignored as insignificant for typical steel messenger applications, for which elastic behavior is assumed; see Section 8.4.

8.3 Sag–Tension Relationship

Based on the small slope condition for the catenary, the following formula results, which indicates that the tension is inversely proportional to the peak sag, when

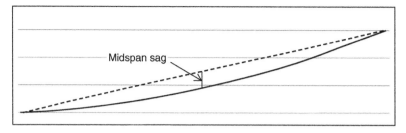

Figure 8.2 Non-level slope.

subject to uniform load, including its own weight.

$$\text{tension (lbs)} = \left[\left(\text{wire weight} + \text{load}\right)\ (\text{lbs/ft})\right]$$
$$\times \left[\text{span length (ft)}\right]^2 / \left[8 \times \text{midspan sag (ft)}\right]$$

where the peak sag is measured from the chord as shown in Figure 8.2, at half the span length, in the direction of the distributed resultant load, and is not necessarily in a vertical direction (see Figure 6.5). In principle, the tension refers to its horizontal component, but the small slope assumption obviates the need in most practical distribution applications for distinguishing between the horizontal component and the total tension, which varies slightly along the span.

As an example, consider a 200 ft span of a relatively heavy (possibly ice covered) cable of 0.5 lbs/ft, with a 2% (=4 ft midspan) sag. The tension would then be calculated as 625 lbs, as follows:

$$\text{tension (lbs)} = 0.5\,\text{lbs/ft} \times \left[200\,\text{ft}\right]^2 / \left[8 \times 4\,\text{ft}\right] = 625\,\text{lbs}$$

For a 1% sag, the tension would be doubled, or 1250 lbs.

8.4 Determining Change in Sag (and Tension)

Changes in sag from initial (installation) conditions may occur because of subsequent changes in mechanical loads or temperature, or support movement. These effects may be quantified by consideration of the amount of "slack" (i.e. difference between the total length along the catenary and the span length, or chord length if non-level slope of support points), as again based on the parabolic approximation:

$$\text{slack (ft)} = (8/3)\ \left[\text{midspan sag (ft)}\right]^2 / \text{span length (ft)}$$

A characteristic of shallow catenaries, which are typical of most practical wire and cable applications, is that the midspan sag is very sensitive to possible movement of the end supports, where the latter are not rigid or fixed. Thus, a relatively

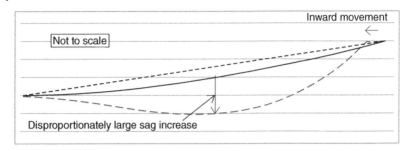

Figure 8.3 Inward movement of support point(s).

small inward (longitudinal) movement of either of the support points will result in a disproportionately large increase in sag. This effect is illustrated in Figure 8.3 and may be closely approximated by the following formula:

$$[\text{subsequent midspan sag (ft)}]^2 = [\text{initial midspan sag (ft)}]^2 + (3/8) [\text{net inward movement (ft)}] \times \text{span length (ft)}$$

in which the "net inward movement" is the arithmetic sum of the horizontal movement of the support points, for which an inward movement is considered positive (+) and an outward movement is negative (−). This formula is based on equating the total length of the wire along the curve after movement to the total length of the wire prior to the movement of the support point(s). This equation assumes that the decrease in tension corresponding to the increase in sag, as indicated by the sag–tension relationship of Section 8.3, is not accompanied by a significant recovery in the strain. Such a reduction in strain would tend to offset the increase predicted by the above formula. This effect may be temporarily ignored for the purpose of understanding the impact of relatively small movements of the support point(s) on the peak sag.

For example, the subsequent sag resulting from a total 12-in. (1 ft) inward movement for the 200 ft span of Section 8.3, and initial 2% (=4 ft) sag, would be estimated as follows:

$$[\text{subsequent midspan sag (ft)}]^2 = [4\text{ft}]^2 + (3/8) [1.0\text{ ft}] \times 200\text{ ft}$$
$$= 91.0\text{ ft}^2$$

corresponding to a subsequent sag of possibly as much as 9½ ft, more than double the initial sag. This 5½ ft increase in sag is more than five times the magnitude of the inward movement itself and is a result of the highly nonlinear nature of the geometric relationship. Whereas in principle, the above formula may be used to determine the decrease in sag resulting from a net *outward* movement – corresponding to a negative (−) net *inward* quantity – of the support point(s), the accompanying increase in tension may be large and the corresponding increase in strain (i.e. increase in wire length) cannot be ignored. Indeed, it may be seen that, for a small outward (negative) movement of only a few inches, the above formula indicates the resulting sag corresponds to the square root of a negative number, which is not meaningful.

Thus, the quantitative change in sag under more general conditions, including possible outward movement, as well as under storm loading, is considerably more complicated than that provided by the above formula. The following algebraic equation determines the subsequent peak (midspan) sag, following an increase in distributed (uniform) loading on the wire, including that due to additional weight (e.g. ice) and/or transverse wind, and temperature change, as well as an inward (longitudinal) movement of the end points.

$$\Delta_2^3 - \left[\Delta_1^2 + (3/8)\,\delta l + (3/8)\,\Delta_{temp}\eta l^2 - (3/64)\left(w_1/EA\right) l^4/\Delta_1 \right]\Delta_2$$
$$- \left[(3/64)\left(w_2/EA\right) l^4 \right] = 0$$

where

Δ_1	=	Initial midspan sag, ft
Δ_2	=	Subsequent midspan sag, ft
δ	=	Net (specified) inward movement of support point(s), ft
ℓ	=	Horizontal (projected) span length, ft
w_1	=	Initial resultant distributed load on wire or supporting messenger, lbs/ft
w_2	=	Subsequent resultant distributed load on wire or supporting messenger, lbs/ft
Δ_{temp}	=	Change in temperature of wire or supporting messenger from initial condition (1) to subsequent condition (2), °F
η	=	Thermal expansion coefficient of wire or supporting messenger, (in./in.)/°F
EA	=	Effective stiffness of wire or supporting messenger, lbs/(in./in.)

The respective sags are measured, from the chord, in the direction of the distributed resultant load, similar to that indicated in Figure 6.5, and are not necessarily in a vertical direction. This equation assumes a linear elastic behavior of the wire or supporting messenger, which is characteristic of steel messenger strands. (Typical values of the stiffness of the messenger strand are provided in **Appendix A**.) However, since supply cables often include components (e.g. aluminum) that exhibit inelastic deformation, this equation would only apply to such cables in their "final" state, after inelastic stretching has occurred (see Section 8.2), and the cable is behaving elastically, within a limited tension. Commercial software tools are available that also include the effects of such inelastic behavior, but which may not consider explicit movement of the support points for an individual span (see Section 8.5).

Although complicated, the above cubic equation for Δ_2 has an exact ("closed form") solution, as provided in **Appendix D**. Quantitative results from this method are consistent with that presented in Figures 7.3 and 7.4, as based on the use of the indicated commercially available software tool, with any small differences essentially attributable to the particular values used for the various parameters shown in Appendix A.

Based on the resulting subsequent sag, Δ_2, the corresponding subsequent tension, T_2 (lbs), may be determined from the sag–tension relation of Section 8.3, or

$$T_2 = w_2 \, \ell^2 / \left(8 \Delta_2 \right)$$

Similarly, the initial tension, T_1 (lbs), corresponds to

$$T_1 = w_1 \, \ell^2 / \left(8 \Delta_1 \right)$$

Alternatively, the above cubic equation may be reformulated in terms of the tensions T_1 and T_2, instead of the sags Δ_1 and Δ_2, as indicated, to directly determine the subsequent tension T_2, using the same mathematical procedure as in Appendix D.

8.5 Ruling Span

The formulas in Section 8.4 allow for the possibility of an explicit (known) movement of the support points in a given span. This effect may occur in many practical situations and is the basis of the "ruling span" method for determining sags and tensions for lines with unequal spans. This method is most applicable to transmission conductors that are individually suspended on free-swinging insulators, for which the movements of the support points result in essentially equal tensions among the individual spans, which are assumed to be located between a section of (rigid) dead-ends. The ruling span is the effective span length, ℓ_{eff}, used to determine the subsequent tension resulting from additional loadings or temperature changes, which are assumed to be uniformly applied across the entire section of individual spans. The effective, or ruling span, distance is given by

$$\ell_{\text{eff}} = \sqrt{\left[\left(\ell_1{}^3 + \ell_2{}^3 + \ell_3{}^3 + \cdots + \ell_n{}^3 \right) / \left(\ell_1 + \ell_2 + \ell_3 + \cdots + \ell_n \right) \right]}$$

where there are "n" individual spans between the dead-ends. This effective length may be significantly greater than the average span length and is determined by equating the total length along the individual spans, including slack, to that of the ruling span, replicated as necessary to equal the total section length between dead-ends. This procedure provides a solution accounting for the increased wire length associated with tension and temperature changes, consistent with the total wire length along the spans, as based on the parabolic model for the catenary.

The ruling span would be used as the span length, ℓ, in the cubic equation of Section 8.4 to determine the (fictitious) sag in the ruling span, where the support movement term, δ, is ignored in this case. This sag is then used to determine the actual subsequent tension along the section, using the sag–tension relationship above, and for which the actual subsequent sag in each individual span may then be determined using the same tension value, again using the sag–tension relationship.

Since the ruling span concept is based on tension equalization among the spans, it may not be valid for less flexible supports than that equivalent to the free-swinging insulators supporting transmission lines. Thus, although individual conductors secured to poles, using posts or clamps may allow some support movement, consistent with that of the stiffness of the pole or support hardware, the assumption of tension equalization may not be valid for typical distribution systems. This may be the case, for example, for an individual cable clamped to the same pole as a group of multiple lines, all with different physical characteristics, and/or mounted at different times, for which the support movements, and individual tension equalization, would be inhibited by the combination of lines acting on the pole.

8.6   Point Load

Although a uniform distributed load is typically used to determine the sag and tension in suspended wires, it is sometimes useful or necessary to consider the effect of a more localized (e.g. point) load in addition to an existing or initial distributed load. The point load may be deliberately caused by a marker ball placed on a supply line, or by communications equipment mounted on a messenger, or accidentally caused by a very heavy item, such as a large fallen branch or a leaning tree. The appropriate method of analysis depends on the magnitude and location of the point load, since this will affect the ability to obtain reasonable estimates of the response of the wire.

Figure 8.4 illustrates two possible responses, for vertical sag, to an asymmetrically located vertical point load (e.g. weight). In Figure 8.4a, the peak sag occurs away from the load, which may be of relatively low magnitude and/or mounted relatively close to a support point (e.g. strand-mounted antenna). Figure 8.4b corresponds to a load of sufficient magnitude such that the peak sag occurs at the point load. In the former case, the added load may actually reduce the peak sag

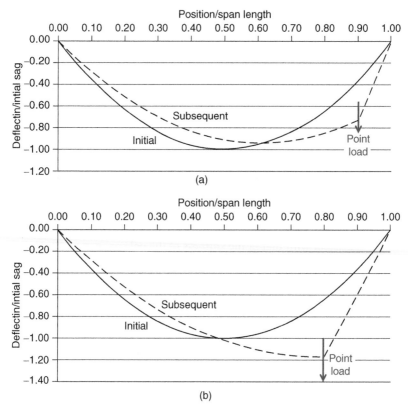

Figure 8.4 (a) Point load – peak sag occurs away from load. (b) Point load – peak sag occurs at load.

that existed prior to the load application, as illustrated, and may be a productive practice for some applications, as necessary.

The results in Figure 8.4, which are shown not-to-scale, are again based on the parabolic model and the assumed relatively shallow slopes of the catenary curve. However, for loads that may be grossly heavier than that of the suspended wire or cable, such as a fallen branch or leaning tree, and/or very close to a support point, the slope may be relatively high in portions of the span, for which a more accurate solution may be desired. Figure 8.5 models the span with two straight line segments, intersecting at the point load, assuming the effect of the point load greatly exceeds that of the initial distributed weight. Unlike the parabolic model, with low slopes, for which the tension is assumed to be essentially constant along the span, the tensions on both sides of the point load in Figure 8.5 are considered to be different.

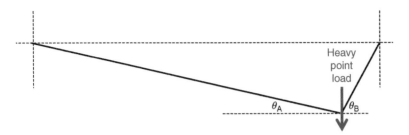

Figure 8.5 Very heavy point load.

The procedures for determining the mathematical solution for both these models, which include the effects considered in Section 8.4 (i.e. elastic stretching of wire, thermal expansion/contraction, movement of support points, etc.) are relatively complicated and is presented in **Appendix E**. The solution determines the peak sag, which is generally not at the midpoint of the span.

The tensions are again inversely related to the peak sag, which in this case would generally result in an asymmetric tension applied to the pole, acting toward the span with the point load, resulting in a net load on the pole in the longitudinal direction. This net load, however, would be reduced by tension equalization resulting from pole flexure, as the various wire(s) sags and tensions respond to such flexure, with the wires tending to balance, or relieve, much of the differential longitudinal load imposed upon the pole itself.

9

General Order 95 (California)

9.1 General Order 95 (GO 95)

Although the National Electrical Safety Code (NESC) is required or recognized as the effective safety standard in most of the United States (see Section 6.1), California has adopted its own requirements in General Order 95 (CPUC 2020). While qualitatively similar in many respects, there are significant quantitative differences in the technical requirements. In addition, as indicated in its purpose, GO 95 also attempts to ensure "adequate service," which is a reliability issue, whereas the primary purpose of the NESC is to protect people from direct physical harm (e.g. electrocution) from the electrical supply and communications facilities.

> **11 Purpose of Rules (GO 95)**
> The purpose of these rules is to formulate, for the ***State of California***, requirements for overhead line design, construction, and maintenance, the application of which will ensure ***adequate service*** and secure safety to persons engaged in the construction, maintenance, operation, or use of overhead lines and to the public in general. [emphasis added]

9.2 Loading Requirements

GO 95 provides a loading district map that is analogous to that of the NESC, as shown in Figure 9.1, with ice and wind loads specified in Table 9.1. In this case, the Heavy loading, with $\frac{1}{2}$-in. radial ice, corresponds to a wind pressure of 6 psf, in comparison to the 4 psf of the NESC Heavy (or Medium) loading, while the Light loading corresponds to a wind pressure of 8 psf, in comparison to the 9 psf of the NESC Light loading. The temperature for determining wire tensions is again 0 °F for the Heavy loading district, but somewhat lower for the Light loading district (i.e. 25 °F for GO 95 vs. 30 °F for the NESC). However, the relative severity

Overhead Distribution Lines: Design and Application, First Edition. Lawrence M. Slavin.
© 2021 The Institute of Electrical and Electronics Engineers, Inc.
Published 2021 by John Wiley & Sons, Inc.

Loading districts

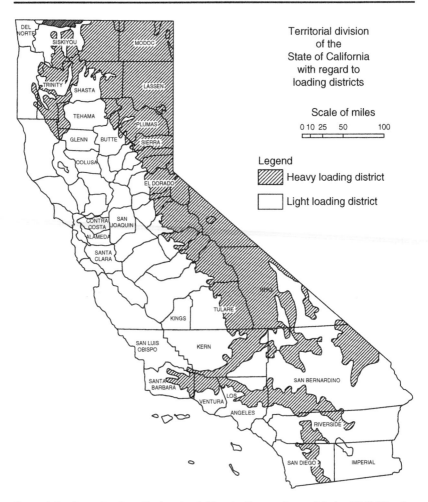

Territorial division
of the
State of California
with regard to
loading districts

Scale of miles

0 10 25   50   100

Legend

▨ Heavy loading district

☐ Light loading district

Figure 9.1   Storm loading districts for California. Source: General Order 95, California Public Utilities Commission.

of the load requirements for the design and strength of the structural and various line components in GO 95 must also be based on consideration of the other specified strength requirements, as discussed in Section 9.3, which tend to result in more conservative designs than typically that of the NESC. In general, many of the principles for the loadings discussed in Section 6.2, regarding the NESC, remain applicable to the GO 95 loadings.

Table 9.1 GO 95 storm loadings.

	District	
	Heavy	**Light**
Ice thickness (in.)	½	0
Wind pressure (psf)	6	8
Temperature (°F)	0	25

9.3 Strength Requirements

For GO 95, the requirement that the structure or other component be able to withstand the imposed loads may be expressed by the relation

Strength/Safety factor ≥ Effect of Load

This is different than the load and resistance factor design (LRFD) format used in the NESC, as explained in Section 6.3, for which the "safety factor" may be interpreted as equivalent to the ratio of the load factor divided by the strength factor, and which is generally required to be ≥1.0. The safety factors in GO 95 are specified as a function of the type component and material, as well as the Grade of Construction.

The definitions of Grades of Construction in GO 95 (Grade A, B, or C) do not match those of the NESC (Grades B, C, and N), especially for joint distribution applications. Typical joint usage in GO 95 requires its highest grade of construction (Grade A), in contrast to that of the NESC for which its highest grade (Grade B) is generally only required for distribution applications (joint or non-joint) for lines crossing railroads, limited-access highways, and navigable waterways requiring crossing permits. Thus, the GO 95 safety factor for the commonly employed joint-use unguyed wooden poles is 4.0, at initial construction, almost double that of the effective "safety factor" of approximately 2-to-1, based on the ratio of the NESC load factor to the strength factor, as explained in Section 6.3 (i.e. 1.75/0.85 = 2.06), for similar (NESC Grade C) applications. In practice, the strength differences are less than double in the Light loading districts, partially because the NESC specifies a 9 psf wind pressure in comparison to the 8 psf pressure in GO 95. The more significant consideration, however, is the 1/3 reduction in the GO 95 safety factor allowed to account for possible deterioration and/or subsequent attachments, prior to "replacement." The NESC does allow a 1/3 reduction for deterioration for wood poles, but does not allow the reduction to apply to additional attachments; i.e. the initial strength requirements continue to apply for the pole in its undeteriorated condition. Thus, for an undeteriorated pole, the equivalent or comparable

GO 95 safety factor, for subsequent additions to a joint-use wooden pole in the Light loading district, is effectively equal to $4.0 \times 8/9 \times 2/3 = 2.37$, which remains greater (more conservative) than the 2.06 effective safety factor of the NESC.

The degree of conservativeness of GO 95 for typical wood poles is even more apparent in the Northeast portion of California for which the GO 95 Heavy storm loads apply, in comparison to the NESC for which the Medium loading is applicable. In this area, the GO 95 wind pressure is 50% greater than, and the ice thickness is double, that of the NESC. The strength requirement for wood poles in this area is therefore more than double that of the NESC, even with the 1/3 allowed reduction, especially for smaller conductors or wires, for which the larger ice loading will be particularly significant.

In contrast to the wooden poles, the GO 95 strength requirements for metal (e.g. steel) pole applications are much lower than that of the NESC in the important GO 95 Light loading area of California, although the requirements may be greater than that of the NESC in the GO 95 Heavy loading areas of California.

In general, a broad comparison between GO 95 and the NESC for various structural elements is complicated by the use of its various safety factors, as opposed to the LRFD format and load and strength factors of the NESC, and for which the appropriateness of the individual safety factors is not necessarily evident. It is anticipated that the California Public Utilities Commission will eventually consider revisions to GO 95 and adopt the LRFD format in the future.

9.4 Clearances

Similar to the NESC, GO 95 specifies minimum clearances of wires and conductors above surfaces, such as roads and buildings, as well as minimum clearances between wires. The method of determining or measuring the appropriate clearances is, however, significantly different. Whereas the NESC generally requires the clearances to be maintained under "worst-case" conditions, such as with ice loading or high temperature, GO 95 specifies the clearances to be maintained at a nominal temperature of 60 °F. The quantitative values in GO 95 are therefore greater than those of the NESC, in order to allow for inevitable decrease when subject to differential ice loads or temperature conditions, but for which GO 95 only allows a 5–10% reduction. This, for example, corresponds to significantly greater clearances in GO 95 for primary power over roads. An important category is the clearance between supply and communications conductors, referred to as the communication worker safety zone (CWSZ) in the NESC (Section 7.4), and for which the GO 95 separation tends to be considerably more conservative.

10

Examples

10.1 Purpose

The presented examples illustrate the application of the strength and loading rules, as well as a means of estimating the potential impact of an additional attachment under consideration, without necessarily requiring a formal analysis, or using a sophisticated software tool. Furthermore, the ability to perform sample calculations, using a spreadsheet, for relatively straightforward cases, will enable the user of such software tools to gain experience and an understanding of the various issues, allowing a sanity check on the generated results. Such expertise will help the user avoid blindly accepting obvious erroneous results, created by incorrect inputs or other user errors that may have not otherwise been apparent.

Consider a 40-ft long, Class 4 wood pole, in the Heavy loading district, with 200 ft spans on both sides, supporting three supply conductors, approximately ½-in. diameter each, mounted toward the top of the pole, and two communications lines, approximately 1-in. diameter each, including the 6M messenger (**Appendix A**), mounted half-way up the pole; see Figure 10.1. These specifications are deliberately somewhat vague, since the purpose is only to estimate the additional loads on the pole system, in comparison to its capacity and available margin. Typical Grade C construction requirements are assumed.

10.2 Tangent Line

For a tangent line, the primary loads of interest on the structure are the wind forces tending to bend the pole in a transverse direction, as a cantilever, with the critical load generally corresponding to the groundline bending moment, or the equivalent effective force ("class load") applied near (i.e. two feet from) the top of the pole (Section 2.2, Table 2.1). It is noted that the wind forces on the conductors (ice-covered, as appropriate), as transferred to the structure, account for the

Overhead Distribution Lines: Design and Application, First Edition. Lawrence M. Slavin.
© 2021 The Institute of Electrical and Electronics Engineers, Inc.
Published 2021 by John Wiley & Sons, Inc.

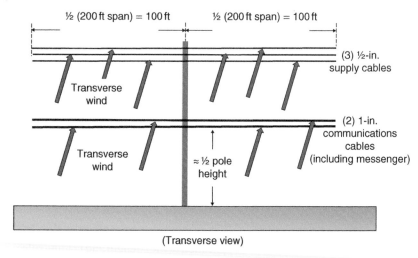

Figure 10.1 Example pole with supply and communications lines.

dominant forces applied to the pole and are generally much greater than the wind force on the pole itself, or on support hardware or auxiliary items, for which ice loading is not necessarily added. Nonetheless, it would be appropriate for such additional wind forces to be included in the calculations for any formal analysis intended to provide a baseline for any pole of interest, which may be accomplished by a spreadsheet type analysis (RUS 2014), for which an available software tool may also possibly be employed.

Assuming the pole is embedded at 6 ft, the supply conductors would be mounted at a height of approximately 32–34 ft and the communications cables at approximately 17 ft. (The precise heights will obviously vary, based on appropriate spacings and ground clearance requirements.) The diameters of the ($\frac{1}{2}$-in. radial) ice-covered supply and communication wires are $1\frac{1}{2}$-in. (= 1/8 ft) and 2-in. (= 1/6 ft), respectively. The corresponding wind forces transferred to the pole are equal to the diameters multiplied by the average span length (= 1/2 of 200 ft on each side of the pole), for each cable or line, and then multiplied by the specified wind pressure of 4 psf.

In order to determine the effective cantilever load relative to the class load applied near the top of the pole, these forces are then reduced in proportion to the height of application relative to that near the top of the pole. For the present purposes, the forces on the supply cables will not be adjusted for their precise height of application (vs. 2 ft from the top of the pole), but the forces on the communications cables must be approximately halved. Thus, the effective transverse load is calculated as

$$3 \ \text{(supply cables)} \times 1/8 \ \text{ft} \ \text{(diameter)} \times 200 \ \text{ft} \ \left(\text{average span}\right)$$
$$\times \ 4 \ \text{lbs/ft}^2 \ \text{(wind pressure)} + 2 \ \text{(communications lines)} \times 1/6 \ \text{ft} \ \text{(diameter)}$$
$$\times \ 200 \ \text{ft} \ \left(\text{average span}\right) \times 4 \ \text{lbs/ft}^2 \ \text{(wind pressure)} \times 1/2$$
$$\left(\text{fraction of height of pole}\right) = 300 \ \text{lbs} \ \text{(supply)} + 1/2$$
$$\times \ 267 \ \text{lbs} \ \text{(communications)}$$
$$\approx 435 \ \text{lbs}$$

This force must be multiplied by the appropriate load factor, which is equal to 1.75 for typical (Grade C) distribution applications, resulting in an effective specified load of approximately **760 lbs**. In comparison, the allowable (i.e. permitted) cantilever load for a Class 4 pole is equal to its average strength (2400 lbs) multiplied by the corresponding strength factor of 0.85, or 2040 lbs, for a utilization of less than 40% capacity. Thus, while only approximate in nature, these calculations provide useful information on the status of the pole, with regard to its strength and loading conditions. It is recognized that the wind loads on the pole, including possible equipment, will somewhat reduce the remaining margin, but will likely not affect the overall conclusion regarding the general status of the pole.

Although this example indicates that the previously commonly used "4/40" (Class 4, 40 ft long) pole is more than sufficient for a relatively long span, the proliferation of communications companies and advanced technologies has led to crowded poles, and the need for longer, stronger poles, such as a "3/45" (Class 3, 45 ft long), or possibly larger size pole.

In this example, the load contribution of the supply cables significantly exceeds that of the communication lines. For the same configuration in the Light loading district, in the absence of ice, the relative contributions may be somewhat different. The cable diameters subject to the wind pressure are then $1/2$-in. (=1/24 ft) and 1-in. (=1/12 ft) for the supply and communication lines, respectively, and the effective transverse load is then given by

$$3 \ \text{(supply cables)} \times 1/24 \ \text{ft} \ \text{(diameter)} \times 200 \ \text{ft} \ \left(\text{average span}\right)$$
$$\times \ 9 \ \text{lbs/ft}^2 \ \text{(wind pressure)} + 2 \ \text{(communications cables)}$$
$$\times \ 1/12 \ \text{ft} \ \text{(diameter)} \times 200 \ \text{ft} \ \left(\text{average span}\right) \times 9 \ \text{lbs/ft}^2 \ \text{(wind pressure)}$$
$$\times \ 1/2 \ \left(\text{fraction of height of pole}\right) = 225 \ \text{lbs} \ \text{(supply)}$$
$$+ \ 1/2 \times 300 \ \text{lbs} \ \text{(communications)}$$
$$\approx 375 \ \text{lbs}$$

While the relative contribution of the communications lines to the supply lines is somewhat greater, the net loading, including load factor, is less than that in the Heavy district, corresponding to a utilization of approximately 35% or less. Nonetheless, as stated above, the expanding number of communications attachments will tend to require longer, stronger poles in the future.

10.3 Line Angle

Consider the same application as above, but where the pole is at a corner with a "pull" of 25 ft (Figure 2.8), or a total angle of almost 30°; see Figure 10.2. (In this example, the indicated 100-ft distance in Figure 10.2 is equal to half the 200 ft span length, as well as the reference length which determines the "pull." In general, of course, half the span length would not coincide with the reference length.) The supply conductors are assumed to have a rated breaking strength (RBS) of 5000 lbs, with the (6M) communications messengers having a rated strength of 6000 lbs. The wire tensions contribute a significant transverse load on the pole, as may be determined by the formula in Section 2.5, in addition to that from the wind loads above. The determination of the tensile loads under storm conditions (ice, wind, temperature) is a nontrivial calculation, as explained in Section 8.4. A convenient alternative that is sometimes used, although often very conservative, is to assume the wire tensions are at their 60% limit specified in the NESC for Grade C (or Grade B) construction – i.e. 3000 lbs (supply) and 3600 lbs (communications).

In this case, the additional (effective) transverse loads on the pole may be conservatively estimated as

$$3 \ \text{(supply)} \times 3000 \ \text{lbs (wire tension)} \times 25 \ \text{ft} \ \text{(pull)} \ /50 \ \text{ft}$$

$$+ \ 2 \ \text{(communications)} \times 3600 \ \text{lbs} \ \text{(wire tension)} \times 25 \ \text{ft} \ \text{(pull)} \ /50 \ \text{ft}$$

$$\times \ 1/2 \ \left(\text{fraction of height of pole}\right)$$

$$= 4500 \ \text{lbs} \ \text{(supply)} + 1/2 \times 3600 \ \text{lbs} \ \text{(communications)}$$

$$= 6300 \ \text{lbs}$$

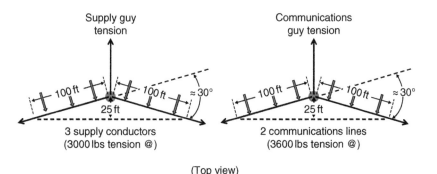

(Top view)

Figure 10.2   Example pole at line angle supported by guys.

This load must be multiplied by the load factor for wire tensions, as applied to the wood pole, which is equal to 1.30 for a Grade C application (Table 6.2), resulting in a required load of approximately **8200 lbs**.

This load is an order of magnitude greater than that caused by the wind pressure, and several times that of the strength of the pole in question (Class 4), as well as that of much stronger poles, even if at half of the assumed storm tension. Therefore, guys would typically be employed, which are assumed to withstand the full applied load for the wood pole application. Ideally, such guy wires would be attached at the pole at the same levels of the supported loads, to avoid local bending effects in the pole. In practice, however, it is usually sufficient to use one guy to support closely spaced supply lines and another guy for closely spaced communications lines, although additional guys are sometimes used. The (single) supply guy would then have to balance the 4500 lbs due to wire tension plus the 300 lbs wind load (Heavy loading district), while the (single) communication guy would have to balance the 3600 lbs plus the 267 lbs wind load, acting at that level. (Although the transverse component of the wind load on the wires will be somewhat lower due to the line angle, it is conveniently assumed to the same magnitude as for the tangent structure.) These loads, however, must be multiplied by the appropriate load factors, which are different for the tension (=1.10 for wire tensions applied to supporting guys) and wind pressure (=1.75); see Table 6.2 and appropriate footnotes. Thus, the corresponding tension in the supply guy is based on an unbalanced load given by

$$= 4500 \text{ lbs} \times 1.10 \quad \text{(wire tension)} + 300 \text{ lbs} \times 1.75 \text{ (wind load)}$$
$$\approx 5500 \text{ lbs}$$

For an assumed lead/height ratio of $\frac{1}{2}$ (Section 2.5), the tension in the supply guy is then calculated as

$$= 5500 \text{ lbs} \times \left[\sqrt{\left(1 + \left(\tfrac{1}{2} \right)^2 \right)} \right] / \left(\tfrac{1}{2} \right)$$
$$= 12\ 300 \text{ lbs}$$

Considering the strength factor (=0.9) for guys, a strand, or combination of strands, equivalent to 16M would be more than sufficient.

The similar calculation for the communication guy, for the same lead/height ratio ($\frac{1}{2}$), gives an unbalanced load

$$= 3600 \text{ lbs} \times 1.10 \quad \text{(wire tension)} + 267 \text{ lbs} \times 1.75 \quad \text{(wind load)}$$
$$\approx 4400 \text{ lbs}$$

with a corresponding guy tension of almost 10 000 lbs, for which a 10M strand, including the strength factor (0.9) would be acceptable, considering its actual strength of 11 000 lbs (see Appendix A) and the conservative assumption of the

6M messenger being loaded to the NESC limit of 60% of its rated strength. This selection is not as strong as that recommended in the Telcordia *Blue Book* for the corresponding messengers (6M), pull (25 ft), and lead/height ratio, but this guideline is apparently based on a guy that can withstand the full breaking strength of the messenger.

The required strengths of the anchors are not directly considered in this analysis, but may be similarly addressed, recognizing that the load and strength factors applicable to the anchors (and soil foundation) may be different than that used for the guy wires. In particular, Table 6.2 indicates the load factor for the metal portions of the anchor will again be 1.10, as for the guy wire, but the load factor for the (soil) foundation will be the default 1.30, while the strength factors for the anchor and soil foundation will be 1.0.

10.4 Line Angle – Buckling Consideration

Since the wooden pole is assumed to be fully supported by the guy system, it may appear that the pole strength (Class) is not of importance. However, the large vertical forces imposed on the pole by the guys (Figure 10.3) may be sufficient to cause the pole to buckle.

For this analysis, which considers the vertical load applied to the wood pole, and its corresponding resistance (strength) to buckling, different load and strength factors are applicable for the pole than for the guy wires. Thus, the load factor of 1.90 applies to vertical loads on wood structures, and a strength factor of 0.85

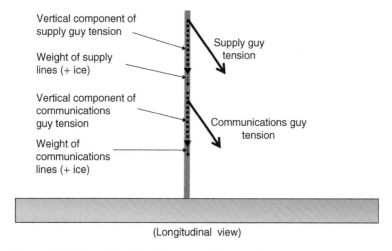

(Longitudinal view)

Figure 10.3 Example pole subject to vertical loads from guy tension.

applies to the buckling strength. For this purpose, the unfactored wire tensions and wind forces are used to determine the unbalanced load, which then determines the guy wire tension, and corresponding vertical load, to which the load factor of 1.90 is applied. The unbalanced load, without load factors, for the supply lines is therefore

$$= 4500 \text{ lbs} \quad (\text{wire tension}) + 300 \text{ lbs} \quad (\text{wind load})$$

$$= 4800 \text{ lbs}$$

for which the corresponding tension in the supply guy is then calculated as

$$= 4800 \text{ lbs} \times \left[\sqrt{\left(1 + (\tfrac{1}{2})^2\right)} \right] / (\tfrac{1}{2})$$

$$\approx 10\,700 \text{ lbs}$$

Based on the formula in Section 2.5, the vertical component is then

$$= 10\,700 \text{ lbs} / \sqrt{\left(1 + (\tfrac{1}{2})^2\right)}$$

$$\approx 9600 \text{ lbs}$$

resulting in a factored vertical load

$$= 9600 \text{ lbs} \times 1.90 \quad (\text{vertical})$$

$$\approx \mathbf{18\,200 \ lbs}$$

The similar calculation for the communication guy gives an (unfactored) unbalanced load

$$= 3600 \text{ lbs} \quad (\text{wire tension}) + 267 \text{ lbs} \quad (\text{wind load})$$

$$\approx 3900 \text{ lbs}$$

and a guy tension

$$= 3900 \text{ lbs} \times \left[\sqrt{\left(1 + (\tfrac{1}{2})^2\right)} \right] / (\tfrac{1}{2})$$

$$\approx 8700 \text{ lbs}$$

with a vertical component

$$= 8700 \text{ lbs} / \sqrt{\left(1 + (\tfrac{1}{2})^2\right)}$$

$$\approx 7800 \text{ lbs}$$

resulting in a factored vertical load

$$= 7800 \text{ lbs} \times 1.90 \quad (\text{vertical})$$

$$\approx \mathbf{14\,800 \ lbs}$$

These vertical loads are very high and will generally dominate any additional loads from the supported conductors and equipment, ice-covered, as appropriate. For example, assume the three $\tfrac{1}{2}$-in. supply conductors each weigh $\tfrac{1}{4}$ lbs/ft, and the two 1-in. (including messenger) communications cables each weigh $\tfrac{1}{2}$ lbs/ft.

Using the formula in Section 6.2 for the ice-covered weight (Heavy loading district) of a supply cable gives a unit weight of

$\frac{1}{4}$ lb/ft (cable weight)

$+ 1.24 \times \frac{1}{2}$ (ice thickness, in.) $\times [\frac{1}{2}$ (cable diameter, in.)

$+ \frac{1}{2}$ (ice thickness, in.)]

≈ 0.9 lbs/ft

or a total weight, for the three supply cables, for a 200 ft span, of

3×0.9 lbs/ft $\times 200$ ft

$= 540$ lbs

The unit weight for each ice-covered communication line is given by

$\frac{1}{2}$ lb/ft (cable weight)

$+ 1.24 \times \frac{1}{2}$ (ice thickness, in.) $\times [1$ (cable diameter, in.)

$+ \frac{1}{2}$ (ice thickness, in.)]

≈ 1.4 lbs/ft

or a total weight, for the two communications cables of

2×1.4 lbs/ft $\times 200$ ft

$= 560$ lbs

These weights, even with the load factor applied, are only a very small fraction of the vertical load due to the guy tensions. With the possible exception of heavy transformers, it may therefore be assumed that such additional weight, including that of miscellaneous equipment, as well as the pole itself, may be ignored for the purpose of estimating the potential for pole buckling, due to the uncertainties in estimating the dominating wire tensions, and especially the buckling resistance.

The RUS version of the Gere and Carter method, described in Section 2.6, may be conveniently employed to estimate the buckling resistance. In this method, the vertical loads are added and the total is assumed to be applied at the lower load (communication guy) point. For the present example, the total load of approximately **33 000 lbs** (=18 200 + 14 800 lbs) is applied half way up the pole. The following formula is then used to determine the buckling load:

$$W_{bu} = n\pi^2 EI_{eff}/h_{eff}^2$$

for which

$n = 1$
$E = 2\,130\,000$ lbs/in.^{2} (assume Southern Pine, ANSI 2017)
$h_{eff} = 17$ ft
$I_{eff} = I_w \times (D_{GL}/D_w)^2$

and I_w is the cross-section moment of inertia at the lower guy load, and is given by

$$I_w = \pi D_w^4/64$$

The ANSI O5.1 standard specifies the following (minimum) circumferences for the 4/40 pole (Southern Pine) at 6 ft from the butt, C_6, and the pole top, C_{top}, as well as the "approximate distance from groundline" (i.e. embedment depth), GL_{dist}:

C_6 = 33.5 in.

C_{top} = 21 in.

GL_{dist} = 6 ft

for which the "taper" may therefore be calculated as

---

taper   = $(C_6 - C_{top})/(\text{length} - 6)$

            =(33.5 − 21)/(40 − 6) in./ft

            =0.37 in./ft.[1]

Based on the taper, and the assumption that the actual embedment depth is equal to the approximate value given by GL_{dist} (6 ft), the pole circumferences, C_w and C_{GL}, may now be calculated at the lower guy location and groundline, respectively:

$$C_w = C_6\text{-taper}\left(h_{eff} + \text{depth–}6\right)$$
$$= 33.5\text{–}0.37\,(17 + 6\text{–}6)$$
$$= 27.2 \text{ in.}$$

$$C_{GL} = C_6\text{-taper (depth–}6)$$
$$= 33.5\text{–}0.37\,(6\text{–}6)$$
$$= 33.5 \text{ in.}$$

The above calculation for the groundline circumference (C_{GL}) results in a value equal to that specified 6 ft from the butt (C_6), since GL_{dist} and the assumed embedment depth are also equal to 6 ft. However, the method shown is general for other poles and/or embedment depths.

The relevant moments of inertia I_w and I_{eff} may now be calculated as follows:

$$I_w = \pi\left(C_w/\pi\right)^4/64$$

---

1 Alternatively, a different taper value is indicated in the Annex (Informative –nonmandatory) of ANSI O5.1, which may be used for determining the circumference at the groundline. The corresponding circumference would be slightly different, but not consequential for the present purposes of estimating the bucking load.

$$= \pi \ (27.2/\pi)^4 /64$$
$$= 276 \ \text{in.}^4$$

$$I_{\text{eff}} = I_{\text{w}} \times \left(D_{\text{GL}}/D_{\text{w}}\right)^2$$
$$= 276 \times \left[\left(C_{\text{GL}}/\pi\right)/C_{\text{w}}/\pi\right]^2$$
$$= 276 \times \left[33.5/27.2\right]^2$$
$$= 419 \ \text{in.}^4$$

Thus,

$$W_{\text{bu}} = n\pi^2 E I_{\text{eff}}/h_{\text{eff}}^2$$
$$= (1) \ \pi^2 \left(2\,130\,000 \ \text{lbs/in}^2\right) \left(419 \ \text{in.}^4\right)/(17 \ \text{ft} \times 12 \ \text{in.}/\text{ft})^2$$
$$\approx 200\,000 \ \text{lbs}$$

or, **170 000 lbs** (=0.85 × 200 000) with the strength factor applied. Thus, the buckling resistance is significantly greater than the total applied load of **33 000 lbs**, indicating that potential buckling is not of concern.

10.5 Additional Attachment

A question that commonly arises is the feasibility of a communications service provider to install an additional cable required to meet the growing needs of its customers. In general, there are two possible alternatives for adding another line, including (i) the placement of a new strand and cable, as usually appropriate for a new third-party attacher, and (ii) overlashing the new cable to an existing lashed cable(s) bundle, as appropriate for an already existing communications attacher. The required additional pole strength may be estimated, without necessarily resorting to sophisticated software tools, using similar principles as those for evaluating the pole status with respect to the supply and communications lines discussed above. This incremental strength may then be compared to the present utilization, or available margin, to verify feasibility, assuming such a baseline or status has been previously established, possibly using commercially available software tools. The pole owner(s) may make educated decisions as to when new detailed pole evaluations may be warranted, based on a sequence of such incremental increases.

Assuming space is available for a new strand, consistent with clearance requirements (Chapter 7), consider the first option of adding a new strand to support the new required communications cable. Similar to above, a new 1-in. diameter bundle, including 6M messenger, plus ½-in. radial ice (=2-in., or 1/6 ft, ice-covered), is considered, mounted at approximately mid-height of the Class 4 pole, with 200 ft. spans. The incremental storm load is then calculated for the one (additional) communication line, or

$$1 \text{ (communications line)} \times 1/6 \text{ ft (diameter)} \times 200 \text{ ft (average span)}$$
$$\times 4 \text{ lbs/ft}^2 \text{ (wind pressure)} \times 1/2 \text{ (fraction of height of pole)}$$
$$= 1/2 \times 133 \text{ lbs}$$
$$\approx 65 \text{ lbs}$$

This force must again be multiplied by the appropriate load factor of 1.75, resulting in an effective load of approximately 115 lbs, or somewhat more than 5% of the permitted 2040 lbs cantilever load for the Class 4 pole. This is well within the available margin for the discussed tangent line, and not of immediate concern.

However, for the case of the line angle, the additional transverse pull placed on the pole by the new strand has a more significant effect on the pole support system, including the supporting guy wire connected at the communications level. For the present purpose of estimating the impact of the additional attachment on the guy and anchor system, it is sufficient to only consider the effect of the messenger tension, which is much greater than that due to the wind load on the ice-covered wire, as seen above (Section 10.3). Thus, again using the convenient, albeit conservative, assumption that the messenger strand will be limited to 60% of its rated strength (3600 lbs = 0.60 × 6000 lbs), and accounting for the load factor of 1.10 for wire tension, and the amplification of more than double corresponding to the lead/height ratio of ½, results in an incremental guy/anchor load of approximately

$$1 \text{ (communications line)} \times 3600 \text{ lbs (wire tension)} \times 25 \text{ ft (pull)} / 50 \text{ ft}$$
$$\times 1.10 \text{ (load factor)} \times \left[\sqrt{\left(1 + (1/2)^2\right)} \right] / (1/2) \text{ (lead/height amplification)}$$
$$\approx 4400 \text{ lbs}$$

This load, in addition to the 10 000 lbs for the original two communications lines (Section 10.3), and accounting for the 0.9 strength factor, indicates a 16M strand guy would now be necessary.

Regarding the increased possibility of pole buckling, the wide margin between the vertical loads resulting from the messenger tensions, including appropriate load factors, and the estimated resistance to buckling, including strength factor, indicates this additional vertical load would not be of concern.

While the above estimates indicate the incremental loads from the proposed additional communications line, on its own new messenger, are not very consequential, it is nonetheless instructive to compare these incremental loads to that corresponding to the overlash alternative. The efficiencies and benefits of overlashing are discussed in detail in Sections 5.3, 6.8, and 7.5, with sample quantitative results illustrated in Figures 6.7 and 6.8 regarding incremental loads and tensions under storm conditions. In the present example, for the tangent line, the following modified calculation applies to the incremental transverse wind load on the ice-covered overlashed bundle:

$$1 \text{ (communications bundle)} \times \left[\Delta_{OD}/12\right] \text{ ft (incremental diameter)}$$
$$\times 200 \text{ ft (average span)} \times 4 \text{ lbs/ft}^2 \text{ (wind pressure)}$$
$$\times 1/2 \text{ (fraction of height of pole)}$$
$$= 1/2 \times 800 \times \left[\Delta_{OD}/12\right] \text{ lbs}$$

where the term Δ_{OD} refers to the incremental diameter (inches) of the bundle, with or without ice. While the bundle size, with ice, is larger than that of the original ice-covered bundle, it is much less than that of a separate new messenger and cable, with ice. For the assumed initial 1-in. bundle, an incremental diameter Δ_{OD} of $1/2$-in. or less would be anticipated; e.g. see Figure 5.7. The incremental load would then be

$$= 1/2 \times 800 \times \left[0.5/12\right] \text{ lbs}$$
$$\approx 15 \text{ lbs}$$

Including the load factor of 1.75, the effective incremental load is less than 30 lbs, or slightly more than 1% of the permitted 2040 lb cantilever load for the Class 4 pole, much less than the already small incremental load for the separate messenger, and essentially negligible.

In the case of the line angle requiring the guy support system, the same conservative assumption that the messenger is at the NESC limit of 60% rated strength would in all likelihood be equally applicable after the addition of the second cable to the strand, as suggested by the results illustrated in Figure 6.8. (In practice, this condition should be verified, possibly by comparison with other situations for which tensions have been established, or possibly by other calculations, including the method of Section 8.4, or equivalent commercially available tool.) Thus, the original guy system should still be sufficient following the overlash, and the ability of the pole to resist buckling, caused by the downward (vertical) component of the guy tensions, would not require further consideration.

10.6 Summary

The examples illustrate that it is often not difficult to determine the general status of a pole, with regard to its strength in comparison to the loading requirements, or the implications of an additional attachment, without the need to utilize sophisticated software tools. Such a determination may be more problematic in the presence of significant line angles because of the load contributions of the wire tension, which determination is a separate issue (Section 8.4). However, conservative assumptions may be made, such as assuming the wire tensions are at the NESC limit of 60% RBS, as required for Grade C (or Grade B) construction, facilitating the analysis, but may lead to unnecessarily strong guy wire systems. An additional complication in such cases is that the buckling strength of the pole may need to be estimated to verify the pole can withstand the potentially large vertical loads that may be imposed by such guy systems. On the other hand, for such a designed system, the addition of a new line by overlashing to an existing bundle, may not require any additional strength-related considerations, other than that the strand tension is still considered to be within the 60% limit.

The example in Section 10.3 assumes a relatively large line angle for which support guys are obviously required. Guying is not necessarily required for relatively small line angles, which may be consistent with the ability of the pole to support itself. Some guidelines suggest that poles with a pull less than 2–3 ft (i.e. total line angle approximately 2½ degrees) may not require guying (Lucent 1996). Nonetheless, except for very small angles, the corresponding transverse load resulting from the wire (storm) tensions may be a major factor and must be considered in determining the proper pole size. For very small angles (e.g. a pull less than 1 ft), the combination of the low angle and pole flexure tending to further reduce the angle, as well as the conservative design practices (load and strength factors, etc.), should allow the transverse contribution of the wire tensions to be ignored for most practical applications.

Appendix A

Properties of Messenger Strands

Strand (designation)	Diameter (in.)	Weight (lbs/ft)	Breaking strength (lbs)	Thermal coefficient (in./in./°F)	Effective stiffness (lbs/in./in.)
2.2M[a]	3/16	0.077	2 400	0.000 006 4	443 100
6M	5/16	0.225	6 000	0.000 006 4	1 278 900
6.6M[b]	1/4	0.121	6 600	0.000 006 4	847 600
10M	3/8	0.270	11 000	0.000 006 4	1 581 300
16M	7/16	0.390	16 000	0.000 006 4	2 295 300
25M	1/2	0.510	25 000	0.000 006 4	3 028 200

a) Limited usage (drops, etc.)
b) Recent limited usage vs. 6M strand.
Source: Suggested values shown, based on historical telephone industry information (AT&T 1952; Telcordia 2017); specific values may vary, depending upon source.

Overhead Distribution Lines: Design and Application, First Edition. Lawrence M. Slavin.
© 2021 The Institute of Electrical and Electronics Engineers, Inc.
Published 2021 by John Wiley & Sons, Inc.

Appendix B

Wireless Attachments

The need or desire to install wireless facilities on utility poles has led to antennas typically mounted at the very top, sometimes using a pole-top extension, adding several feet to the total height. This location is often available and may be optimum with respect to maximizing coverage, depending on the wireless technology employed, but presents complications for installation and maintenance because of its position relative to supply voltages. Future installations for small cells may also require crossarm or sidearm mounting, as well as strand-mounting by which an antenna may be supported on an existing (or new) messenger in the communications space. The lower height of the strand-mount, for example, avoids issues related to proximity to supply lines, and may also be more appropriate for efficient transmission and reception for some networks. Figure 4.13 illustrates pole top, sidearm/crossarm and strand-mounted locations. The various installation issues are discussed in the Telcordia *Blue Book*, including the clearance requirements of the NESC, and those related to the protection of persons from the effects of radiation (Telcordia 2017).

An interesting aspect of the option of using a pole-top extension is the suggestion that the effectively greater height of the pole should require a greater depth of embedment, consistent with the greater depths required for naturally longer poles (Section 2.4), generally corresponding to a greater anticipated overturning moment at the groundline. Such a concept would be difficult to implement, if valid, since it would require re-installation of the structure, or appropriate remediation to increase the stability of an existing pole. Ironically, however, the artificially taller pole is at least as stable (regarding resistance to overturning) as the original structure since the groundline bending moment must be limited to that consistent with its GL dimensions, as discussed in Section 2.2. Thus, the overall horizontal (cantilever) forces must be proportionately reduced such as to maintain, or not exceed, the characteristic GL bending (i.e. overturning) moment. Furthermore, models for estimating the allowable overturning moment, such as the RUS method (Section 2.4), predict a slightly greater stability, which is apparently due

Overhead Distribution Lines: Design and Application, First Edition. Lawrence M. Slavin.
© 2021 The Institute of Electrical and Electronics Engineers, Inc.
Published 2021 by John Wiley & Sons, Inc.

to the beneficial effect of the correspondingly lower shear forces at the ground level. The addition of a pole-top extension is therefore not an issue with regard to pole stability, with no need to increase an otherwise acceptable embedment depth.

In contrast, there are potential clearance issues for strand-mounted antennas, similar to other such mounted equipment. Since the minimum clearances to surfaces below are usually determined at or near the midpoint of the span, due to wire sag, the clearances will generally not be reduced for equipment mounted close to the pole, as is typically the case. Indeed, it is quite possible the midspan clearance will actually increase due the overall response of the wire to a local (point) load, such as illustrated in Figure 8.4a. Rather, the most likely problem will be reduced clearances between communications lines of different utilities mounted in relatively close vertical proximity, primarily because of the physical dimensions of the equipment, possibly violating the 4-in. minimum clearance between their facilities. In general, the variability of communications attachments, and the frequency of changes, including the need to serve subscribers, as well as to introduce new technologies (5G, etc.), presents a challenge to the practical ability to universally maintain the specified clearances between communications lines.

The addition of any facilities increases the load on the pole, including during the storm conditions specified by the NESC. For tangent lines, the critical forces are primarily those resulting from the horizontal wind pressure acting on the exposed (projected) cross-sectional areas, for which the load contributions of the wireless facilities will be proportional to their size (dimensions), as well as their attachment height. The relevant projected areas include that of the antenna and auxiliary equipment, but especially any supporting cables spanning the poles; i.e. the wind force on the cable is often the major contribution, significantly exceeding that on the equipment itself. However, if the associated cable(s) may be overlashed to an already existing cable–messenger system, the corresponding incremental load may be very low (Section 6.8). Thus, the net increase in loading because of the addition of wireless facilities will vary, depending upon the required cables and equipment, and how installed, as well as the existing pole class, span length and loading district – sometimes representing only a few percent of the pole capacity, and possibly less in some cases, although the quantitative estimates may depend upon the assumptions inherent in the theoretical analysis.

For structures at line angles, the wire tensions are important regarding the strength of the pole system, including guys, in which case any significant increase in weight due to a strand-mounted object could conceivably be of importance. Since, however, the antennas and/or associated equipment would typically be mounted close to the pole, any such incremental tensions tend to be relatively low, as may be verified employing the method described in Section E.1 for the addition of a point load.

Appendix C

Extreme Wind and Extreme Ice Loadings

Table C.1 indicates the loadings corresponding to the extreme wind and extreme ice (with concurrent wind) of Rule 250C and Rule 250D, respectively. (The additive constant is equal to 0.0 for these cases.) In comparison to the three distinct loading districts of Rule 250B, these loads are provided by contour maps, for which the various wind speeds and/or ice thicknesses may vary within geographic areas, including individual states. The maps are provided by the American Society of Civil Engineers.

These loads are applied in the same manner as described in Chapter 6 for the district loads, but for which the applicable load and strength factors are those in Table C.2.

Similar to the allowed 33% reduction of the required strength of wood poles to account for deterioration when subject to the Rule 250B loads (Section 6.6), a 25% reduction is applicable for the Rule 250C and 250D loads.

A significant difference between the district load criteria and that of the extreme winds, is the need to calculate the wind pressures corresponding to the wind speeds indicated in NESC Figures 250-2 (Extreme wind) and 250-3 (Extreme ice). For Rule 250D, the appropriate wind pressures may be determined by the simple formula of Section 6.6:

$$\text{Wind pressure } (\text{lbs/ft}^2) = 0.00256 \times (\text{Wind speed})^2 \times \text{Shape factor}$$

which, with an assumed shape factor of 1.0, results in the values (to 1 decimal place) in Table C.3.

In contrast, the determination of the wind pressure for Rule 250C is considerably more complicated, and includes the use, and possible calculation, of additional parameters, as in the following equation:

$$\text{Wind pressure } (\text{lbs/ft}^2) = 0.00256 \times (V_{3\,\text{sec}})^2 \times k_z \times G_{\text{RF}} \times C_f$$

Overhead Distribution Lines: Design and Application, First Edition. Lawrence M. Slavin.
© 2021 The Institute of Electrical and Electronics Engineers, Inc.
Published 2021 by John Wiley & Sons, Inc.

where

V_{3sec}	=	3 second wind speed
k_z	=	Velocity pressure exposure coefficient
G_{RF}	=	Gust response factor
C_f	=	Shape factor

The terms k_z and G_{RF} are determined for the structure or wire (or component) by equations provided in the NESC, which are relatively complicated, and for which tabulated values are therefore also provided (NESC 2017). However, for typical distribution lines, the value of the k_z term for the pole itself varies between 0.9 and 1.0, while typical values for the wire (or component) vary between 1.0 and

Table C.1   Rule 250C and 250D storm loadings.

	250C	250D
Ice thickness (in.)	0	See NESC Figures 250-3
Wind speeds (mph)	See NESC Figures 250-2	See NESC Figures 250-3
Temperature (°F)	60	15

Table C.2   NESC load and strength factors for Rule 250C and 250D extreme loads.

		Grade B	Grade C
Load factors	Rule 250C Extreme wind	1.00	**0.87**[a]
	Wind loads		
	All other loads	1.00	1.00
	Rule 250D Extreme ice with concurrent wind	1.00	1.00[b]
Strength factors	Wood[c]	0.75	**0.75**
	Metal[d]	1.0	1.0
	Support hardware	0.8	0.8
	Guy wire	0.9	0.9
	Guy anchor and foundation	1.0	1.0

a)   For wind velocities above 100 mph (except Alaska), a factor of 0.75 may be used.
b)   The radial ice thickness shall be multiplied by a factor of 0.80.
c)   Also includes reinforced concrete.
d)   Also includes pre-stressed concrete and fiber-reinforced polymer.

Table C.3 Rule 250D pressure vs. wind speed.

Wind speed (mph)	Pressure (lbs/ft²)
30	2.3
40	4.1
50	6.4
60	9.2
70	12.5
80	16.4

1.1, depending upon the relevant height, with the slightly lower values applying at lower height. The G_{RF} term is approximately 1.0 for the pole (or component), and 0.9 for wires, for typical distribution lines. The shape factor, C_f, is assumed equal to 1.0 for cylindrical shapes (e.g. wires and poles) and 1.6 for flat surfaces. (The same shape factor applies to application of the wind pressures of Rule 250D and the Rule 250B district loads of Section 6.2.)

In contrast to the maximum allowable tension of 60% of the rated breaking strength under the Rule 250B district loading (Section 6.4), a tension of 80% of the rated breaking tension is allowed for the extreme loads of Rules 250C and 250D.

Appendix D

Solution of Cubic Equation

The algebraic cubic equation in Section 8.4 for determining the subsequent sag, Δ_2, resulting from a change in the loading condition, may be solved as described below.[1]

The equation is of the form $z^3 + a_2 z^2 + a_1 z + a_0 = 0$, where "$z$" corresponds to the subsequent (mathematically unknown) sag Δ_2, and coefficients a_0, a_1, and a_2 are constants (i.e. mathematically known quantities). In this case, the term "a_2" $= 0$, and "a_1" and "a_0" are the negatives of the corresponding terms in the brackets [ ] of the cubic equation. The solution depends on the sign (positive or negative) of the quantity Q defined as

$$Q \equiv q^3 + r^2$$

where

$$q = a_1/3 \quad \text{and} \quad r = -a_0/2$$

It may be verified, for the sag equation in Section 8.4, the term r is positive (>0) while the term q may be positive (>0) or negative (<0), depending on the effects or loads imposed on the wire. Therefore, the quantity Q may be positive or negative.

For $Q > 0$, the solution of interest is the sum of the two terms:

$$s_1 = \left(r + \sqrt{Q}\right)^{1/3}$$

$$s_2 = \left(r - \sqrt{Q}\right)^{1/3}$$

or

$$z = \left(r + \sqrt{Q}\right)^{1/3} + \left(r - \sqrt{Q}\right)^{1/3}$$

For $Q < 0$, the above solution is also applicable, in spite of the included square root of a negative quantity ($\sqrt{Q}$). In this case, the solution must be mathematically

---

1 *Handbook of Mathematical Functions*, M. Abramowitz and I.A. Stegun, Dover Publications, Inc., 1970.

Overhead Distribution Lines: Design and Application, First Edition. Lawrence M. Slavin.
© 2021 The Institute of Electrical and Electronics Engineers, Inc.
Published 2021 by John Wiley & Sons, Inc.

expressed in terms of complex variables, including real and imaginary components. The quantities s_1 and s_2 may then be expanded as

$$s_1 = \left[\sqrt{(r^2 - Q)}\right]^{1/3} \cos\left\{(\text{atan}\left[(\sqrt{-Q}\,)/r\right])/3\right\} + i\left[\sqrt{(r^2 - Q)}\right]^{1/3} \times \sin\left\{(\text{atan}\left[(\sqrt{-Q}\,)/r\right])/3\right\}$$

and

$$s_2 = \left[\sqrt{(r^2 - Q)}\right]^{1/3} \cos\left\{(\text{atan}\left[(\sqrt{-Q}\,)/r\right])/3\right\} - i\left[\sqrt{(r^2 - Q)}\right]^{1/3} \times \sin\left\{(\text{atan}\left[(\sqrt{-Q}\,)/r\right])/3\right\}$$

where "i" is the unit imaginary quantity ($\sqrt{-1}$). The solution of interest is again equal to the sum of the terms s_1 and s_2, which is seen to be

$$z = 2\left[\sqrt{(r^2 - Q)}\right]^{1/3} \cos\left\{(\text{atan}\left[(\sqrt{-Q}\,)/r\right])/3\right\}$$

For $Q = 0$, both solutions for z are equal to $2r^{1/3}$.

Appendix E

Point Load

E.1 Parabolic Model

Figure 8.4 shows separate catenaries on both sides of the vertical point load, each of which may be approximated by separate parabolas, as appropriate for applications where the slope of the curvature is low. The form of the solution depends upon whether the peak sag occurs at the point load or if it occurs elsewhere – i.e. Figure 8.4a (Mode I) or Figure 8.4b (Mode II). Both cases again require the solution of a cubic equation, similar to that of **Appendix D**, but as based on the parameters defined below.

Mode I applies under the following condition:

$$d \geq [(wl + 2P)/(wl + P)]l/2$$

where l is the horizontal (projected) span length, d is the distance from the point of load application to one of the support points and required to be $\geq l/2$ (i.e. greater or equal to half the span length), without any loss of generality, w represents the distributed vertical load on the supporting wire, and P is an added local (vertical) point load. Mode II applies under the opposite condition, i.e.

$$d \leq [(wl + 2P)/(wl + P)]l/2$$

For Mode I, the following terms are defined:

$$F_1 = (w/2)\,[l/2 + P(l - d)/(wl)]^2$$
$$G_1 = \{2(-d^2 t + dt^2 + l^3/3) - (l + d)(l^2 - 2dt) + (l^2 - 2dt)^2/[2(l - d)]\}/t^4$$
$$t = l/2 + P(l - d)/(wl)$$

For Mode II, the following terms are defined:

$$F_{||} = [d(l - d)/l]\,[(wl/2) + P]$$

Overhead Distribution Lines: Design and Application, First Edition. Lawrence M. Slavin.
© 2021 The Institute of Electrical and Electronics Engineers, Inc.
Published 2021 by John Wiley & Sons, Inc.

$$G_{||} = (1/2)\,[-dH_{||}(1 + J_{||}) + (1 + J_{||})^2/d + H_{||}^2\,l^3/3 - (l + d)\,H_{||}(K_{||} - 1)$$
$$+ (K_{||} - 1)^2/(l - d)]$$

$$H_{||} = w/F_{||}$$

$$J_{||} = wd^2/(2F_{||}) = H_{||}d^2/2$$

$$K_{||} = w(l^2 - d^2)/(2F_{||}) = H_{||}(l^2 - d^2)/2$$

Similar to the applications without a point load, the peak sag and tension are inversely related as in Section 8.3, but by the following more general relationship:

$$\text{tension (lbs)} = F(\text{lbs} - \text{ft})/[\text{peak sag (ft)}]$$

where $F = F_I$ or F_{II} for Mode I or Mode II, respectively. Similarly, the slack may be expressed by the following more general relationship

$$\text{slack (ft)} = G(\text{ft}^{-1}) \times [\text{peak sag (ft)}]^2$$

where $G = G_I$ or G_{II} for Mode I or Mode II, respectively.

Using these relationships and functions F and G, an equation analogous to the equation in Section 8.4 may be developed; i.e.

$$\Delta_2{}^3 - [8\Delta_1{}^2/(3lG) + \delta/G + \Delta_{\text{temp}}\eta l/G - wl^3/(8\text{EA }G\Delta_1)]\,\Delta_2 - Fl/(\text{EA }G) = 0$$

for which Δ_1 and Δ_2 represent the peak (vertical) sags, before and after the addition of the point load, respectively, and the other terms are as defined above or in Section 8.4. In this case, however, Δ_2 will generally not occur at the midpoint of the span.

The above procedure may be extended to the more general case for which the initial condition, designated state "1", as well as the subsequent condition, state "2", may have different magnitude distributed loads, and different point loads with respect to their position and/or magnitude. The following generalized equation applies:

$$\Delta_2{}^3 - [(G_1/G_2)\Delta_1{}^2 + \delta/G_2 + \Delta_{\text{temp}}\eta l/G_2 - F_1\,l/(\text{EA }G_2\,\Delta_1)]\,\Delta_2 - F_2\,l/(\text{EA }G_2) = 0$$

In this case, the F_1 and G_1 terms refer to the F and G terms as defined above, at the initial state (1), corresponding to the appropriate mode, while F_2 and G_2 similarly refer to the subsequent state (2). This generalized equation reduces to the previous equation above in the absence of an initial point load and with no change in the distributed load. It also reduces to the equation in Section 8.4, without point loads, but where there is a change in magnitude of the distributed load. These algebraic cubic equations may be solved as described in Appendix D.

Although these complex equations were developed for the support points at the same height or elevation, based on the previous results for a uniform distributed load, it may heuristically be assumed that the solution for the subsequent sag Δ_2 again refers to the distance measured vertically from the chord between support points at somewhat different heights.

The above methodology has been developed assuming both the initial distributed load and the point load act vertically, corresponding to uniformly distributed and local weights. The same procedure, however, may be further extended to estimate the effect, including peak tension, resulting from a concurrent wind load, possibly with ice and/or temperature change. In this case, the initial distributed (vertical) weight on the wire, possibly including ice, would be considered in combination with the horizontal wind load, to obtain a resultant subsequent uniform load, acting at an angle to the vertical, similar to that illustrated in Figure 6.5. The effective point load (e.g. from locally mounted equipment) would be determined by also considering its (vertically oriented) weight in combination with the horizontal wind load acting upon its projected area, to obtain a resultant effective point load. These loads, both the uniformly distributed load and the effective point load, would then be used in the above equation(s) to determine an effective subsequent peak sag, for which the corresponding storm tension may be obtained from the above equation relating tension and peak sag. It is recognized, however, that this procedure would only be an approximation, presumably conservative, since the resultant vectors for the subsequent uniformly distributed load and the point load would not necessarily be parallel, depending upon their weight and cross-sectional areas, including ice for the wire; i.e. the vectors may not be similarly oriented, or acting in a common (inclined) plane.

E.2   Intersecting Straight Lines Model

For cases where the parabolic approximation may not be valid, such as where the wire has a relatively large slope, a more accurate analysis may be performed for the case of a very heavy weight, P, in which two intersecting straight line approximations may be appropriate. Thus, Figure 8.5 illustrates a very heavy vertical load (weight) applied to the stretched wire, at a distance, l_A, from one of the support points ("A") which is required to be $\geq l/2$ (i.e. greater or equal to half the span length), without any loss of generality. For this case, where a "more exact" solution is pursued, both support points are assumed to be at the same elevation.

The final sag at the point load, Δ_2, following some longitudinal movement of the load, ε (defined as positive for movement towards the opposite support point "B") as the wire stretches and the corresponding vector components of the tensions and vertical load equilibrate, as well as some possible (assumed) longitudinal movement of the support(s), may be determined by a relatively complicated procedure. The governing equations are

$$P = \{EA\,[M_A(\Delta_2, \varepsilon) - N_A(\Delta_1, \varepsilon)] + w\,l^2\,l_A/(8\Delta_1)\}$$
$$\times \Delta_2(l - \delta)/[l_A(l_B - \delta_B - \varepsilon)\,M_A(\Delta_2, \varepsilon)]$$

and

$$P = \{EA\,[M_B(\Delta_2, \varepsilon) - N_B(\Delta_1, \varepsilon)] + w\,l^2\,l_B/(8\Delta_1)\}$$
$$\times \Delta_2(l - \delta)/[l_B(l_A - \delta_A + \varepsilon)M_B(\Delta_2, \varepsilon)]$$

which represent two equations in the two unknowns Δ_2 and ε, and for which the following new terms are defined (other terms as defined previously):

$$M_A = \sqrt{[\Delta_2{}^2 + (l_A - \delta_A + \varepsilon)^2]}$$

$$M_B = \sqrt{[\Delta_2{}^2 + (l_B - \delta_B - \varepsilon)^2]}$$

$$N_A = l_A + slack_A$$

$$N_B = l_B + slack_B$$

$$slack_A = 32\Delta_1{}^2(l_A{}^3/3 - l\,l_A{}^2/2 + l^2\,l_A/4)/l^4$$

$$slack_B = 32\Delta_1{}^2(l_B{}^3/3 - l\,l_B{}^2/2 + l^2\,l_B/4)/l^4$$

l_A = Horizontal (projected) distance from support point A to point of load application, prior to any longitudinal movement of the load, ε

l_B = Horizontal (projected) distance from support point B to point of load application, prior to any longitudinal movement of the load, ε; thus, span length $l = l_A + l_B$

δ_A = Inward movement of support point A

δ_B = Inward movement of support point B; net inward movement $\delta = \delta_A + \delta_B$

Because of the intractable nature of the governing equations for solving for the sag Δ_2, as a function of the specified (known) point load, P, it is more convenient to solve for the load P as a function of the sag Δ_2. In this case, the two equations may be shown to be equivalent to first solving for the load movement, ε, as a function of the sag Δ_2, as follows:

$$\varepsilon = (l_B - \delta_B) - \sqrt{([\{[M_A(\Delta_2, \varepsilon) - N_A + w\,l^2\,l_A/(EA\,8\Delta_1)]}}$$
$$\times [l_B\,(l_A - \delta_A + \varepsilon)\,M_B]/[l_A\,(l_B - \delta_B - \varepsilon)\,M_A]\} + N_B - w\,l^2\,l_B/(EA\,8\Delta_1)]^2 - \Delta_2{}^2)$$

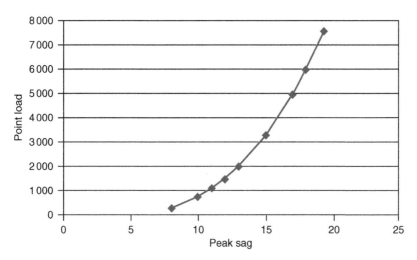

Figure E.1 Point load vs. peak sag.

Although still not tractable, since the term ε is on both sides of the equation, this formulation allows for an iterative technique in which a trial value for the term ε may be inserted into the right side of the equation, and the resulting value of ε on the left side is then used as the next trial value. Depending upon the specific values of the parameters, this procedure will rapidly converge to a solution for ε, which may be inserted into one of the preceding equations to determine the value of the point load, P, corresponding to the selected peak sag Δ_2. This procedure may then be repeated for other selected values for the peak sag, allowing a plot such as hypothetically (without specific units or details) illustrated in Figure E.1. Although indirect and somewhat tedious, this method does provide useful results allowing an understanding of the potential response of the wire or strand to a very heavy load.

The corresponding wire tensions, on both sides of the point load, are then given by

$$T_A = (P/\Delta_2)(l_B - \delta_B - \varepsilon)M_A/(l - \delta)$$

and

$$T_B = (P/\Delta_2)(l_A - \delta_A + \varepsilon)M_B/(l - \delta)$$

where T_A is the tension on the segment of wire towards point A and T_B is the tension in the wire portion toward point B.

Glossary

The following definitions apply when the dictionary definition may not be sufficient for the purposes of this manual.

additive constant the quantity added to the resultant load on a conductor or wire, in addition to the distributed weight and wind force, when determining wire tensions during specified storm conditions.

ADSS an "all-dielectric self-supporting" cable, comprising fibers supported by nonmetallic strength member.

aerial lines outdoor utility lines, including electric power and communications, supported on poles; also referred to as "overhead lines."

anchor system guy wire(s) attached to steel rod (anchor) inserted into the ground, or attached to an essentially stationary object.

angle structure a pole for which the lines extending on opposite sides form an angle significantly different than 180°; see "tangent structure."

belowground plant direct-buried lines or underground conduit system.

bending moment the product of force and perpendicular distance to point of interest, such as the groundline for a pole.

buckling an unstable condition in which the pole collapses, or severely deforms, caused by excessive vertical (axial) load.

butt the very bottom of pole, usually the point with the largest circumference; see "groundline."

catenary the shape of a hanging wire, of uniform weight and negligible bending stiffness, due to gravity; shallow catenaries may be closely approximated by a parabola for most practical applications.

CATV originally used as an acronym for a Cable Television system, a broadband communications system capable of delivering multiple channels of programming, generally by coaxial cable, but also integrating fiber-optic and microwave links; CATV service providers now also offer telephone and Internet services to their subscribers.

Overhead Distribution Lines: Design and Application, First Edition. Lawrence M. Slavin.
© 2021 The Institute of Electrical and Electronics Engineers, Inc.
Published 2021 by John Wiley & Sons, Inc.

circumference the distance (perimeter) measured around the cross-section of the pole, at any designated location; for an assumed circular cross-section, the diameter is equal to the circumference divided by the mathematical quantity π.

clearance the quantitative separation between electric supply or communications facilities and various surfaces, or between the facilities themselves.

clearance zone one of four areas of the United States or territories with specified storm conditions – including radial ice buildup, wind pressure, ambient temperature, and additive constant – for the purpose of determining wire loads that will affect subsequent sags and clearances; see "additive constant", "final sag", and "loading district."

coaxial cable a cable constructed of an inner metallic conductor surrounded by a concentric outer conductor operating at ground potential, capable of transmitting electrical signals for communications purposes, at high frequencies or data rates.

communication worker safety zone (CWSZ) the required minimum (vertical) space between the uppermost communication facility and the lowest supply facility on a joint-use pole, including along the span.

conductor a wire or cable used to transmit electric supply power or electrical communications signals; often used interchangeably with wire or cable.

copper cable a cable comprising individual copper wires, typically arranged in pairs, capable of conducting electrical signals for communications purposes, possibly including associated powering.

dead-end a terminating pole, with lines extending on only one side of the pole.

direct-buried a method of placing cables in trench, in contact with soil, usually without additional covering or protection; see "underground conduit."

distribution outdoor plant facilities providing electric supply or communications service to customers, including lines along local roads and streets, as well as service laterals.

embedment the amount of the pole below the ground level, or the difference between to the pole length and the pole height.

engineering judgment the process of making a decision on the basis of available data and experience, to determine a design or line of action.

feeder outdoor communications facilities, spanning the distance from a central office or headend to a local distribution area, to provide relatively high-capacity communications information.

fiber-optic cable a cable comprising individual optical fibers, typically of glass material, capable of transmitting optical signals for communications purposes, at extremely high data rates; also referred to as "fiber cable."

final sag as used in the NESC, the sag of a wire, including any inelastic deformation resulting from stretching under prior storm loads, or long-term creep at more moderate conditions, when subject to subsequent specified ice loading and/or temperature conditions; see "initial sag."

final tension the tension corresponding to the final sag condition; see "initial tension."

grade of construction a category determining the amount of excess strength required to withstand specified loads.

groundline the base of the aboveground portion of pole; see "butt."

initial sag as used in the NESC, the sag of a wire immediately following installation, and prior to the application of any external load; but also be used in this manual to refer to the amount of sag prior to a change in this value to a subsequent value, for any reason; see "final sag."

initial tension the tension corresponding to the initial sag condition; see "final tension."

joint-use the sharing of structures or facilities, including poles, underground conduit systems, or trenches by several types of service or utilities or organizations, typically by electric power and communications (telephone, CATV) companies.

lead/height ratio the distance from the base of the pole to the point of entry of the guy anchor into the ground, divided by the attachment height of the guy at the pole.

load factor the value that is multiplied by specified loads, to be implemented in the LRFD design procedure; see "load and resistance factor design" and "strength factor."

load and resistance factor design (LRFD) the design procedure in which the specified loads on a structural component are multiplied by appropriate load factors, the effect of which must not exceed the nominal strength of the component multiplied by the specified strength (resistance) factor.

loading district one of four districts of the United States or territories with specified storm conditions – including radial ice buildup, wind pressure, ambient temperature, and additive constant – for the purpose of determining loads on wires and structures, including supported equipment, as appropriate; see "additive constant" and "clearance zone."

messenger a steel wire (strand) used to support cables by the use of lashing wire, rings or hoops, or brackets.

midspan depending upon the context, (i) any portion of the span that is away from the support points, or (ii) the midpoint of the span; see "sag."

modulus of elasticity the property of a material, such as a pole or wire, relating the force required to produce a deformation, for which the material fully recovers when the load is removed.

moment of inertia aka "cross-section moment of inertia", the geometric property of a cross-section that, in combination with the material modulus of elasticity, determines the bending stiffness of a beam, or pole.

nominal strength the resistance of a structural component to a load, usually as provided in an industry standard or document; see "load and resistance factor design."

outside plant outdoor facilities, including structures, equipment, hardware and cables, used by electric power and communications utilities to provide transmission and distribution functions.

overhead lines see "aerial lines."

overlashing the method of lashing an additional cable onto an existing bundle comprising a messenger and previously installed lashed cable(s).

P-delta the additional bending moment resulting from the combination of a significant lateral (horizontal) deflection and a vertical load, usually associated with an overhanging weight.

parabola a mathematical curve that may be used to approximate a shallow catenary; see "catenary."

pole class the classification system that designates the strength of a wooden pole subject to a lateral load applied 2 ft from the top, in a cantilever configuration.

pole height the distance of a pole above the ground, or the pole length minus the depth of embedment; see "pole length."

pole length the distance of a pole from its butt to its tip; see "pole height."

pull at an angle structure, the perpendicular distance of the pole from a chord between points 100 ft from the pole, on the lines on opposite sides of the pole.

primary power relatively high voltage (e.g. 750–22 000 volts) electric power provided along distribution lines, to be converted to low voltage "secondary power" by local transformers, for delivery to customers; see "secondary power."

ruling span the effective span length for a series of poles with unequal span lengths, used to determine resulting sags and tensions for applications with nonrigid (i.e. moveable) support points that tend to equalize the tensions in adjacent spans.

sag the deviation of a suspended wire or conductor, as measured from the chord connecting the two end support points, in the direction of the distributed resultant load, and is not necessarily in a vertical direction; for a uniform weight or load, peak sag occurs at the midpoint of the span; see "midspan," "initial sag," and "final sag."

secondary power electric supply power provided to customers at relatively low voltage; e.g. ≤750 volts; see "primary power."

slack the difference between the total length measured along the catenary of a suspended cable or wire, and the chord length between the support points.

span length the horizontal distance between poles or support points for suspended conductors or wires.

strand see "messenger."

strength factor the value that is multiplied by the nominal strength, to be implemented in the LRFD design procedure; see "load and resistance factor design" and "load factor."

tangent structure a pole for which the lines extending on opposite sides form an angle of approximately 180°; see "angle structure."

tension the force along a conductor or wire (e.g. messenger), determining the corresponding sag.

transmission very high voltage (e.g. >22 000 volts) lines for transporting large power levels to distant locations, for conversion to lower (primary) voltages, at distribution levels, for local applications; also refers to facilities providing long distance communications.

transverse the direction perpendicular to the general direction of the line, in a horizontal plane.

underground conduit a belowground system of pipes allowing supply or communications cables to be conveniently installed within, as necessary; see "direct-buried".

utility a public or private organization providing electric supply or communications services.

wireless the facilities and technologies capable of transmitting communications or other information without the need for physical cables, such as those supporting cellular or mobile service; see "wireline".

wireline the physical cables for transporting electric power or communications, or other information, in contrast to wireless facilities; see "wireless".

References

AASHTO (2011). *Roadside Design Guide*, 4th edition. American Association of State Highway and Transportation Officials.

AFL (2003). *Alumoweld® Type M Guy Strand*. AFL Wire Products.

ANSI (1995). ANSI T1.328-1995, *American National Standard for Telecommunications – Protection of Telecommunications Links from Physical Stress and Radiation Effects and Associated Requirements for DC Power Systems (A Baseline Standard)*.

ANSI (2017). *ANSI O5.1 – Wood Poles: Specifications and Dimensions*. American Wood Protection Association, February 2017.

ASCE (2006). *Reliability-Based Design of Utility Pole Structures, ASCE Manuals and Reports on Engineering Practice No. 111*.

ASCE (2009). *Belowground Pipeline Networks for Utility Cables, ASCE Manuals and Reports on Engineering Practice No. 118*.

ASCE (2019). *Wood Pole Structures for Electrical Transmission Lines, ASCE Manuals and Reports on Engineering Practice No. 141*.

ASCE (2020). *Guidelines for Electrical Transmission Line Structural Loading, 4th edition. ASCE Manuals and Reports on Engineering Practice No. 74*.

AT&T (1952). *Wire and Cable Spans, Wire and Strand Characteristics, Section AO70.10 (919-370-100), Issue 2*. AT&TCo Standard, May 1952.

AT&T (1968). *Erecting Poles and Stubs, Bell System Practices, Section 621-205-200, Issue 2*. AT&TCo Standard, November 1968.

AWPA (2019). *2019 AWPA Book of Standards*. American Wood Protection Association.

CPUC (2017). *General Order No, 165, Inspection Requirements for Electric Distribution and Transmission Facilities*. California Public Utilities Commission, December 2017.

CPUC (2020). *General Order No, 95, Rules for Overhead Electric Line Construction*. California Public Utilities Commission, January 2020.

FHA (1993). *Highway/Utility Guide*. U.S. Department of Transportation, Federal Highway Administration, June 1993.

Overhead Distribution Lines: Design and Application, First Edition. Lawrence M. Slavin.
© 2021 The Institute of Electrical and Electronics Engineers, Inc.
Published 2021 by John Wiley & Sons, Inc.

IEEE (1999). *IEEE P751/D2 (Draft), IEEE Design Guide for Wood Transmission Structures*. IEEE Power Engineering Society.

IEEE (2017). *2017 NESC® Handbook*. Institute of Electrical and Electronics Engineers.

IEEE (2019). *Preprint Proposals for the 2022 Edition of the National Electrical Safety Code (NESC®)*. Institute of Electrical and Electronics Engineers, July 1, 2019.

Lucent (1996). *Outside Plant Systems, Outside Plant Engineering Handbook*. Lucent Technologies, 900-200-318, October 1996.

Marmon (2018). *Hendrix Aerial Cable Systems*, marmonutility.com. Marmon Utility LLC.

Marne, D.J. (2017). *National Electrical Safety Code (NESC®) 2017 Handbook*. McGraw-Hill Education.

NESC (2017). *ANSI C2, National Electrical Safety Code*. Institute of Electrical and Electronics Engineers, August 1, 2016.

RUS (2013). *RUS Bulletin 1730B-121, Wood Pole Inspection and Maintenance*. Rural Utilities Service, United States Department of Agriculture, August 13, 2013.

RUS (2014). *RUS Bulletin 1724E-150, Unguyed Distribution Poles – Strength Requirements*. Rural Utilities Service, United States Department of Agriculture, August 2014.

RUS (2015). *RUS Bulletin 1724E-200, Design Manual for High Voltage Transmission Lines*. Rural Utilities Service, United States Department of Agriculture, December 2015.

RUS (2017). *RUS Bulletin 1724E-205, Design Guide: Preliminary Embedment Depths for Concrete and Steel Poles*. Rural Utilities Service, United States Department of Agriculture, May 2017.

Seiler, J.F. (1932). Effect of depth of embedment on pole stability. *Wood Preserving News* X (11): 152–168, November 1932.

Shoemaker (2017). *The Lineman's and Cableman's Handbook*, 13th edition. (eds. T.M. Shoemaker and J.E. Mack) McGraw-Hill Education.

Southwire (2007). *Overhead Conductor Manual*, 2nd edition. Southwire Company.

Southwire (2009). *18-2 Multiplex Service Drop-Copper*. Southwire, May 29, 2009.

Telcordia (2011). *GR-60, Generic Requirements for Wooden Utility Poles*. Telcordia, December 2011.

Telcordia (2017). *SR-1421, Blue Book – Manual of Construction Procedures*. Telcordia, March 2017; available through Ericsson (https://telecom-info.telcordia.com).

Timoshenko, S.P., J.M. Gere (1961). *Theory of Elastic Stability*, Chapter 4. McGraw-Hill.

Western (2013). *Hendrix Covered Conductor Manual*. Western Power, August 8, 2013.

Index

a

Additive constant *See* National Electrical Safety Code
Aeolian vibration 53
Aluminum 33, 34, 66, 72, 75
Anchors *See* Hardware
ANSI O5.1 8, 15, 26, 49, 57, 93
  brand 11
  butt 8, 93
  circumference 8, 57, 93
  Class 8, 87, 89, 90, 95
  Douglas Fir 8, 49
  fiber strength 8, 49, 57
  groundline 8, 14, 20, 47, 57, 93, 101
  Southern Pine 8, 49, 92
  species 8, 28, 49
  taper 8, 93
  Western Red Cedar 8, 49
Antenna *See* Wireless
ASCE 3, 30

b

Belowground 2, 45
  direct burial 2
  direct-buried 3
  handholes 2
  manholes 2
  underground 2, 45

Brand *See* ANSI O5.1
Buckling 10, 19, 54, 61, 90, 95, 97
  Gere and Carter 21, 92
Butt *See* ANSI O5.1

c

Cable
  ADSS 37
  coaxial 35, 58
  copper pair 36
  fiber 4, 35, 58
  Hendrix 32, 33, 39
  multipair 32
  multiplex 32
Catenary *See* Sag
CATV 4, 37, 63
Circumference *See* ANSI O5.1
Class *See* ANSI O5.1
Clearance 45, 67, 72, 86, 94, 101
  communication worker safety zone 63, 67, 84
  vertical 57, 63, 66, 67
  zone 46
Coaxial *See* Cable
Coefficient of variation 12
Combined ice and wind *See* National Electrical Safety Code
Communication worker safety zone *See* Clearance

Overhead Distribution Lines: Design and Application, First Edition. Lawrence M. Slavin.
© 2021 The Institute of Electrical and Electronics Engineers, Inc.
Published 2021 by John Wiley & Sons, Inc.

Conduit   2
Crossarms *See* Hardware

d

Dead-end *See* Structures
Depth   8, 11, 93, 101
Deterioration   28, 29, 56, 83
  decay   28, 56
  excavation   4, 28
  preservative   11, 26, 30
  zone   28
District *See* National Electrical Safety
    Code
Douglas Fir *See* ANSI O5.1

e

Eccentric   10, 14
Elastic *See* Sag
Embedment   8, 23, 93, 101
  soil   2, 15, 26, 90
Equation   14, 74, 77, 109, 111, 112
  cubic   76, 109
  formula   8, 15, 18, 19, 51, 55, 72, 74,
    91
  iterative   71, 113
Extreme ice with concurrent wind *See*
    National Electrical Safety Code
Extreme wind *See* National Electrical
    Safety Code

f

Feeder   3, 4
Fiber *See* Cable
Fiber strength   8
Formula *See* Equation

g

Gere and Carter *See* Buckling
Grade B *See* National Electrical Safety
    Code

Grade C *See* National Electrical Safety
    Code
Grade N *See* National Electrical Safety
    Code
Guying   20, 31, 97
Guys *See* Hardware

h

Handholes   2
Hardware   7, 51
  anchors   7, 51, 59, 90
  crossarm   7, 23, 38, 39, 46, 49, 101
  ground rods   7
  guys   7, 21, 49, 88
  insulator pins   39
Heavy loading *See* National Electrical
    Safety Code

i

Ice *See* Radial ice
Inspection *See* Maintenance

j

Joint-use   4, 8, 14, 22, 46, 63, 83

l

Lashing   39
  moving reel   39, 40
  overlashing   40
  stationary reel   39
Light loading *See* National Electrical
    Safety Code
Line angle *See* Structures
Load and resistance factor design
  LRFD   49, 83
Load factor *See* National Electrical Safety
    Code
Loads
  lateral   8, 16, 20
  longitudinal   51, 53, 79
  point load   77, 109

pressure   47, 52, 53, 65, 81, 83, 86, 89, 102, 103
transverse   13, 18, 51, 53, 57, 75, 85, 88, 95, 97
vertical   8, 19, 51, 52, 77, 90, 95, 97, 109, 111
weight   10, 51, 53, 58, 71, 73, 75, 77, 92, 102, 111
wire tension   18, 51, 53, 55, 58, 81, 88, 90, 95, 96

m

Maintenance   2, 7, 45, 55, 56, 81, 101
   inspection   26
Manholes *See* Belowground
Medium loading *See* National Electrical Safety Code
Messenger   18, 33, 35, 39, 40, 45, 47, 52, 53, 57, 67, 72, 75, 77, 85, 89, 91, 94, 102
   Alumoweld   33
   strand   18, 23, 38, 58, 89, 95, 97, 101, 113
Modulus of elasticity   19
Moment of inertia   19, 93
Moving reel *See* Lashing

n

National Electrical Safety Code   2, 3, 12, 14, 22, 81
   60 ft   54
   additive constant   47, 52, 65, 103
   Combined ice and wind   54, 56
   district   46, 55, 58, 85, 87, 89, 92, 102
   Extreme ice with concurrent wind   13, 54
   Extreme wind   13, 60
   Grade B   12, 49, 52, 55, 83, 88, 97, 104
   Grade C   12, 50, 52, 83, 85, 87, 88, 97, 104
   Grade N   50, 53
   Heavy loading   46, 58, 85, 89, 92
   Light loading   46, 58, 81, 83, 87
   load factor   49, 51, 52, 83, 87, 88, 90, 95, 104, 117
   Medium loading   46, 58, 84
   Rule 250B   46, 50, 52, 54, 56
   Rule 250C   54, 103
   Rule 250D   54, 103
   strength factor   12, 49, 53, 55, 56, 83, 87, 89, 90, 95, 104
Nonlinear   14, 22, 52, 61, 74

o

Overlashing *See* Lashing

p

Parabola *See* Sag
P-delta   11, 14, 52, 61
Point load *See* Loads
Pole-top *See* Wireless
Preservative *See* Deterioration
Pressure *See* Loads
Primary power   50, 66, 84
Pull   18, 23, 88, 89, 95, 97

r

Radial ice   46, 58, 81, 94, 104
Rated breaking strength   31, 53, 88, 97
Reliability-based design   13
Rule 250B *See* National Electrical Safety Code
Rule 250C *See* National Electrical Safety Code
Rule 250D *See* National Electrical Safety Code
RUS   15, 21, 27, 86, 92, 101

s

Safety factor   52, 83

Sag
  catenary 72, 73, 77, 78
  elastic 72, 75, 79
  final sag 64, 67, 111
  initial sag 74
  parabola 71, 73, 77
Secondary power 50, 53
Shape factor 104
Slack 73, 76, 110
Software 1, 22, 53, 56, 75, 85, 86, 94, 96
Soil *See* Embedment
Southern Pine *See* ANSI O5.1
Span length 31, 73, 76, 86, 88, 102, 109
Species *See* ANSI O5.1
Stationary reel *See* Lashing
Stiffness 49, 53, 59, 61, 64, 71, 75, 77
Storm loading 64, 68, 75, 82
Strand (messenger, guy) *See* Messenger
Strand mounting *See* Wireless
Strength factor *See* National Electrical
     Safety Code
Stress 8, 11
Structures
  dead-end 13, 17, 20, 51, 58, 76
  line angle 13, 14, 17, 20, 49, 53, 58,
     60, 96, 102
  tangent 13, 47, 60, 89, 95, 102

t

Tangent *See* Structures
Taper *See* ANSI O5.1
Telcordia *Blue Book* 18, 22, 41, 51, 89,
     101

Temperature 47, 52, 61, 64, 67, 72, 73,
     76, 81, 84, 88, 104, 111
Tension
  final tension 53, 72
  initial tension 53, 58, 72, 76
Transmission 1, 2, 7, 15, 21, 33, 54, 61,
     76
Transverse *See* Loads

u

Underground *See* Belowground
USDA 28

v

Vegetation 55
Vertical *See* Clearance or Loads

w

Weight *See* Loads
Western Red Cedar *See* ANSI O5.1
Wind speed 47, 103
Wire tension *See* Loads
Wireless 35
  5G 5, 35, 38, 102
  antenna 5, 31, 35, 38, 77, 101
  pole-top 37, 101
  sidearm 37, 101
  strand mounting 37
Wireline 5, 31, 35

z

Zone *See* Clearance or Deterioration

Printed and bound by CPI Group (UK) Ltd, Croydon, CR0 4YY